DigitalMath

Math In Your Hands

Noel Kalicharan

Senior Lecturer, Computer Science
The University of the West Indies
St. Augustine, Trinidad

A publication of the Mental Arithmetic Institute of Trinidad & Tobago

Published August 2012

© Noel Kalicharan, 2012

noel.kalicharan@sta.uwi.edu
digitalmathtt@gmail.com

Preface

Which school subject arouses the strongest emotions in people? You guessed it—mathematics. Some will say flat out that they hate math, others merely dislike it. A few will say they love it—almost certainly, they are good at it. Those who are considered 'good at math' are so few that they are generally regarded as more intelligent or 'have more brains' than the rest of us.

I believe that those who 'hate' math really hate the idea of not being able to do it well. Most likely, they had difficulties early on and these were never resolved. If we try our hand at something and we fail, we don't try so hard the next time, especially if our efforts are ridiculed and discouraged by teachers, parents and peers. So we fail again.

After a little while, we give up, thinking we are no good at this. It is this emotional and mental road-block, *not lack of ability*, that prevents most of us from doing well in mathematics.

Recently, on a whim, I told a group of people I wanted to do a little experiment in mathematics. They all groaned so I asked them to raise their hand if they disliked math. Most did but a young lady raised both hands, signifying her intense dislike, even saying "I hate it!"

I spent the next ten minutes showing them how to multiply by 11 and square numbers that end in 5. She picked it up faster than anyone else. *Within ten minutes*, her whole attitude to math had changed. She wanted to know where she could learn more of these methods. I jokingly replied that the classes were very expensive. She said it didn't matter, she was willing to pay whatever the cost if learning math could be so easy and so much fun.

What this illustrates is that we may *think* we hate mathematics when we really don't. What is lacking is just a little success. When you enjoy little successes, interest grows and makes you want to know more. This breeds more success.

Like it or not, numbers play a big part in our everyday lives and those who are good at numbers enjoy an advantage over those who aren't. One goal of this book is to make you "good at numbers" or, at the very least, better than you were before. And what better way to do this than by using your own two hands?

Welcome to DigitalMath™, a revolutionary way to do arithmetic with your fingers. Have you always believed that you could count to just 10 on your fingers? With DigitalMath, you can count to 99 and beyond. With this ingenious technique, you can do arithmetic with speed and confidence, fully assured that your answers are correct. In many cases, you will be able to calculate faster than someone using a calculator. In fact, the technique is so effective that most abacus-based systems are based on the principles of DigitalMath. Here, your hands play the role of an abacus.

The methods are simple but do not let their apparent simplicity fool you. There is a great deal of mathematical sophistication behind the operations you will perform. And though the methods are easy to learn, it will require a fair amount of practice to master them and make them work for you. Like playing a musical instrument or a sport, it is quite easy to learn the notes of the instrument or the rules of the sport. But it takes years of practice to learn to play well or become good at the sport.

Fortunately, in just a few weeks, an adult can become quite adept at DigitalMath. Small children will take a little longer but can acquire most of the skills in three to six months. The time taken depends mainly on how much time one is willing to devote to practice.

There is no age restriction to learning the techniques of DigitalMath. The only prerequisite is being able to count to 99. Even small children who can count to 20 only can derive many of the benefits. Indeed, we encourage adults to attend the same classes as their children. In these situations, everyone just seems to have more fun.

But the benefits of DigitalMath are not confined to improved arithmetic skills. The use of the hands, in coordination with the brain, stimulates the left and right hemispheres of the brain, releasing creative juices. This leads to whole brain development.

The student who masters these techniques will perform at a higher level in mathematics. As mentioned before, they will be regarded as 'more intelligent' and will be treated differently by teachers and fellow-students. This becomes a self-fulfilling prophecy in that they *will* perform better, not only in mathematics, but in other subjects as well.

No matter what your mathematics achievement so far, even if it is close to zero, I promise that when you see how easy it is to become good at this aspect of mathematics, you will change the way you think of the subject, and this change will bring more success.

My hope is that you will go through the same experience as many of those I've taught, that you will reach the point where you will be awed that math (or at least this part of math) could be so easy and so enjoyable.

Thanks for reading this book. The classes are not very expensive—just the cost of the book—and I trust you will soon learn to love mathematics with both hands.

Noel Kalicharan

Dedicated to the memory of my parents

James & Clara Kalicharan

Contents

$$\boxed{\begin{array}{c} \textbf{1} \\ \textbf{Introduction – the basics} \end{array}}$$

I n this chapter, we introduce all the basic ideas you need to know in order to get started with DigitalMath. We will tell you how fingers get values and how you can represent any number from 0 to 99 using just the ten fingers of your left and right hands.

1.1 Your working position

In order to perform DigitalMath, you need a fairly flat surface—a desk top, table top or even your lap—on which to place your hands. A finger gets a value when it is placed on the surface. We will refer to the surface as the "table".

A good working position is one in which you can rest both hands comfortably on the table. The part of your arm from the elbow to the wrist should be fairly horizontal. More importantly, you must be able to see your fingers easily when you place both hands face down on the table.

Hold your hands a couple inches above the table with all fingers extended but slightly curled. It will *never* be necessary to fold (close) any of your fingers. Your fingers should be extended (unfolded) at all times—this will enable you to calculate faster.

1.2 Names for your fingers

In order to make it easy to refer to a particular finger, we will name them as in the following diagrams.

Right Hand

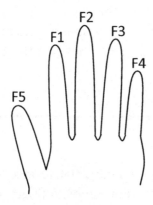

F1 – index, F2 – middle, F3 – ring, F4 – little, F5 – thumb

Left Hand

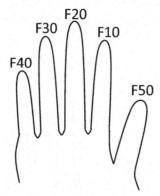

F10 – index, F20 – middle, F30 – ring, F40 – little, F50 – thumb

1.3 How a finger gets a value

In DigitalMath, you use your fingers to store any value from 0 to 99. Each of these values is represented by a unique combination of your ten fingers. (In chapter 12, we will show you how to work with numbers which run into the hundreds.)

The right hand stores the "units" digit and the left hand stores the "tens" digit. For example, if you need to store the number 52, you will store 5 on the left hand and 2 on the right hand. But how do we do that?

First, we'll work with the right hand, keeping in mind that almost everything we say about this hand will apply to the left.

A finger gets a value when it touches the table. Each of the four fingers is worth 1 and the thumb is worth 5.

In the diagrams in this book, we will use a *clear finger* to mean that it is *not* touching the table and a *grey finger* to mean that it *is* touching the table. So the following means that no finger is touching the table

while

means that fingers F1 (index) and F2 (middle) are touching (giving the value 2, see next).

It is not important that the entire finger touch the table—the tip alone is good enough. The only thing that's important is that you can tell the difference between a finger that is touching the table and one that is not.

1.4 Set and clear

We will use the term **set** when we want you to place a finger on the table and **clear** when we want you to remove it. For example **set F1** means to place F1 on the table and **clear F1** means to lift F1 off the table.

✦ ✦

In many of the exercises, you will need to perform some action with your fingers (like setting or clearing) and, at the same time, *say* something. For example, you may have to "set F1" and say "one".

We will use the following convention in this book. Commands like **set** or **clear** will be shown in a box, like this:

set F1   clear F4

and words (or symbols like + and -) to be spoken will be written using this font, for example, 1 or that is. Generally, the command will be placed above the hand and the word(s) or symbol(s) to be spoken will be placed on the lower part of the hand. You perform the command and speak the words/symbols. Some words/symbols to be spoken may not be associated with any command. Such words would be placed on the left or right of the hand. For example, given

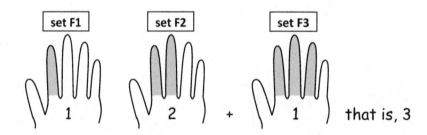

this will be performed as follows:

1. set F1 (say "one")
2. set F2 (say "two")
3. say "plus"
4. set F3 (say "one")
5. say "that is, 3"

1.5 Values for your right-hand fingers

We now show you how numbers are represented on your fingers. We start with the right hand.

+ +

The number 1

You get the value 1 when you "set F1", as shown here:

Setting F1 is the *only* way to store the value 1. It would be *wrong* to set any other finger. For instance, it would be *wrong* to set F2 and say you have 1.

+ +

The number 2

You get the value 2 when you "set F1, F2", as shown here:

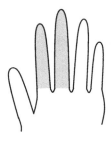

Setting F1, F2 is the *only* way to store the value 2. It would be *wrong* to set any other fingers. For instance, it would be *wrong* to set F3, F4 and say you have 2.

The number 3

You get the value 3 when you "set F1, F2, F3", as shown here:

Setting F1, F2, F3 is the *only* way to store the value 3. It would be *wrong* to set any other fingers. For instance, it would be *wrong* to set F2, F3, F4 and say you have 3.

The number 4

You get the value 4 when you "set F1, F2, F3, F4", as shown here:

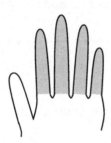

Setting F1, F2, F3, F4 is the *only* way to store the value 4.

The correct order to set and clear

It is very important that you set and clear fingers in the correct order. Fingers must *always* be set in the order F1, F2, F3, F4 and cleared in reverse order (F4, F3, F2, F1).

For example, if you have set F1, F2 and you want to add 1, you *must* set F3. Similarly, if you have set F1, F2, F3, F4 and you want to take away 1, the *only* finger you can clear is F4.

Exercise: Practise counting from 0 to 4 and back to 0: 0-1-2-3-4-3-2-1-0. Repeat.

The number 5

You get the value 5 when you "set F5", as shown here:

Setting F5 is the *only* way to store the value 5.

The number 6

You get the value 6 when you "set F5, F1", as shown here:

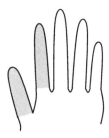

Recall that the thumb is worth 5 and each finger is worth 1. Setting F5, F1 is the *only* way to store the value 6.

The number 7

You get the value 7 when you "set F5, F1, F2", as shown here:

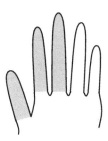

Recall that the thumb is worth 5 and each finger is worth 1. Setting F5, F1, F2 is the *only* way to store the value 7.

The number 8

You get the value 8 when you "set F5, F1, F2, F3", as shown here:

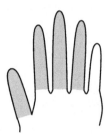

Recall that the thumb is worth 5 and each finger is worth 1. Setting F5, F1, F2, F3 is the *only* way to store the value 8.

<hr>

The number 9

You get the value 9 when you "set F5, F1, F2, F3, F4", as shown here:

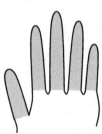

Recall that the thumb is worth 5 and each finger is worth 1. Setting F5, F1, F2, F3, F4 (all the fingers) is the *only* way to store the value 9.

<hr>

Do Exercise 1.1, page 97

1.6 Count from 0 to 9 and back to 0

The following shows how to count from 0 to 9 and back to 0. *Pay particular attention to the movement from 4 to 5 and from 5 to 4.*

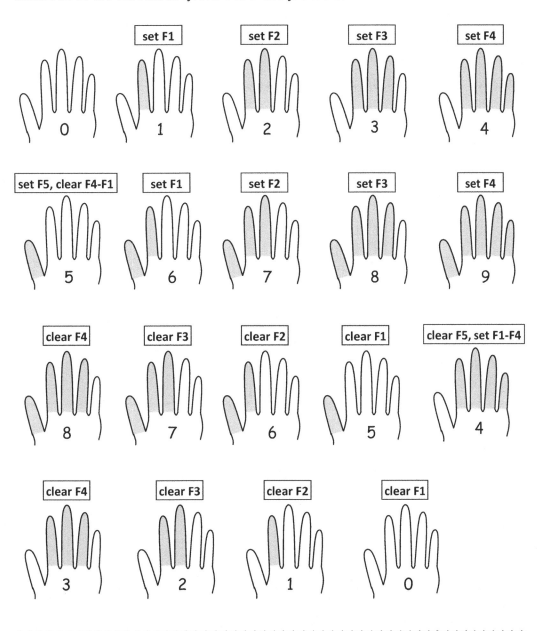

Do Exercise 1.2, page 97

1.7 Values for your left-hand fingers

The left hand is used to store the tens digit of a 2-digit number. For example, if we have to store the number 75, we will store 7 on the left hand and 5 on the right. Just as the tens digit can be a number from 0 to 9, so too the left hand will store a number from 0 to 9, similar to how they are stored on the right hand. But *because* they are stored on the left hand, we multiply by 10 to get their *value*.

+-+

The number 10

You get the value 10 when you "set F10" (left hand), as shown here:

Setting F10 is the *only* way to store the value 10. It would be *wrong* to set any other finger. For instance, it would be *wrong* to set F20 and say you have 10.

+-+

The number 20

You get the value 20 when you "set F10, F20" (left hand), as shown here:

Setting F10, F20 is the *only* way to store the value 20. For instance, it would be *wrong* to set F30, F40 and say you have 20.

The number 30

You get the value 30 when you "set F10, F20, F30" (left hand), as shown here:

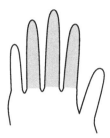

Setting F10, F20, F30 is the *only* way to store the value 30. It would be *wrong* to set any other fingers. For instance, it would be *wrong* to set F20, F30, F40 and say you have 30.

The number 40

You get the value 40 when you "set F10, F20, F30, F40" (left hand), as shown here:

Setting F10, F20, F30, F40 is the *only* way to store the value 40.

The correct order to set and clear

It is very important that you set and clear fingers in the correct order. Fingers must *always* be set in the order F10, F20, F30, F40 and cleared in reverse order (F40, F30, F20, F10). For example, if you have set F10, F20 and you want to add 10, you *must* set F30. Similarly, if you have set F10, F20, F30, F40 and you want to take away 10, the *only* finger you can clear is F40.

Exercise: Count from 0 to 40 and back to 0: 0-10-20-30-40-30-20-10-0. Repeat.

The number 50

You get the value 50 when you "set F50", as shown here:

Setting F50 is the *only* way to store the value 50.

The number 60

You get the value 60 when you "set F50, F10", as shown here:

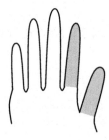

Recall that the thumb is worth 50 and each finger is worth 10. Setting F50, F10 is the *only* way to store the value 60.

The number 70

You get the value 70 when you "set F50, F10, F20", as shown here:

Recall that the thumb is worth 50 and each finger is worth 10. Setting F50, F10, F20 is the *only* way to store the value 70.

The number 80

You get the value 80 when you "set F50, F10, F20, F30", as shown here:

Recall that the thumb is worth 50 and each finger is worth 10. Setting F50, F10, F20, F30 is the *only* way to store the value 80.

+ +

The number 90

You get the value 90 when you "set F50, F10, F20, F30, F40", as shown here:

Recall that the thumb is worth 50 and each finger is worth 10. Setting F50, F10, F20, F30, F40 (all the fingers) is the *only* way to store the value 90.

+ +

Do Exercise 1.3, page 98

1.8 Count from 0 to 90 and back to 0

The following shows the count from 0 to 90 in steps of 10. *Pay particular attention to the movement from 40 to 50 and from 50 to 40.*

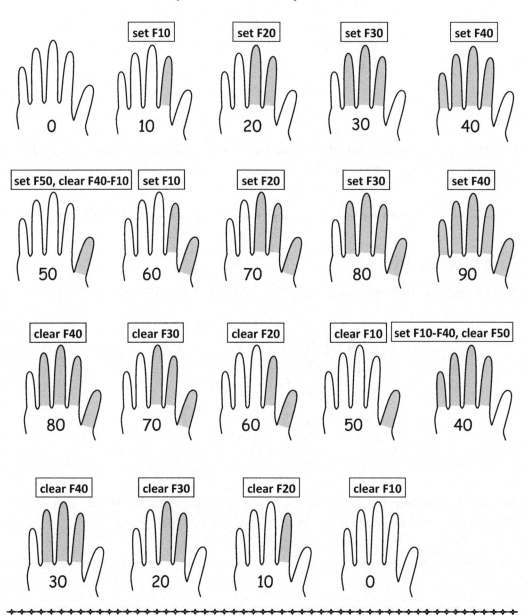

Do Exercise 1.4, page 98

1.9 Values using both hands

As we have seen, we can store the digits 0 to 9 on the left hand and 0 to 9 on the right hand. This means we can store *any* value from 0 to 99 using both hands. For example, to store 75, we store 7 on the left hand and 5 on the right, like this:

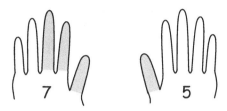

We can count up to 9 on our right hand. To count more, we must use our left hand. Here we show how to count from 7 to 10. Notice the big change from 9 to 10. But it is no different from how we write. We are accustomed to writing "10" after "9". Here we do the same thing except we use our hands as the paper.

Let's see how to count from 27 to 30:

48 to 51:

and 83 to 86:

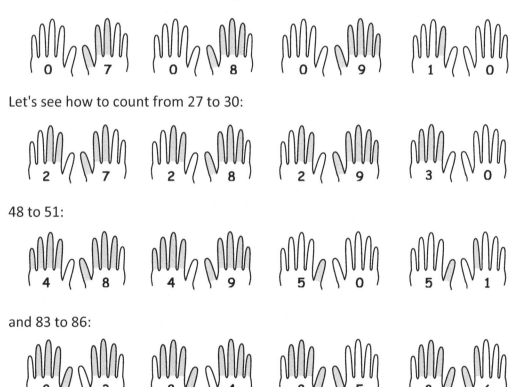

 Do Exercise 1.5, page 99

2
Basic addition/subtraction on right hand

Now that we have shown you how values can be represented using your fingers, it's time to do some arithmetic. Naturally, the first problems will be quite simple (like 2 + 2) and for which you would obviously know the answer. But we caution you to follow the instructions carefully since you will be laying the foundation to do more complex problems later on.

The best advice we can give you is to *let the answers form on your fingers*. This will happen if you follow the steps we describe. Try not to let your knowledge of arithmetic get in the way.

For instance, we all know that 2 + 2 = 4. But you will not gain anything if you simply say the answer is 4 and set 4 on your right hand. However, you will take the first baby steps correctly if you count to 2, say "plus" then count another 2. After this, your hand should tell you that the answer is 4. (This problem is explained on the next page.)

With these remarks, let us start our journey into DigitalMath—the fast, easy and fun way to do arithmetic on your fingers.

2.1 Addition with fingers F1–F4

In the following, do not set 2 or 3 in one movement; *count* until you get 2 or 3. Remember that you add, in order, from the index finger to the little finger.

$$1 + 2 = 3$$

We do this as follows:

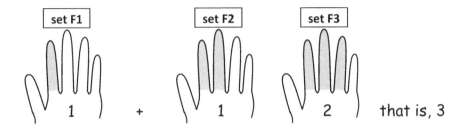

that is, 3

As a reminder, you should say, "One plus one two, that is 3".

$$2 + 2 = 4$$

We do this as follows:

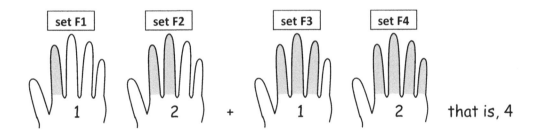

that is, 4

As a reminder, you should say, "One two plus one two, that is 4".

Exercise 2.1

1. 1 + 1
2. 1 + 3
3. 2 + 1
4. 2 + 2
5. 3 + 1
6. 1 + 2 + 1
7. 2 + 1 + 1
8. 1 + 1 + 2

2.2 Subtraction with fingers F1-F4

In the following, do not set 2, 3 or 4 in one movement; *count* until you get 2, 3 or 4. When minusing, take away one at a time. Remember that you *add*, in order, *from the index finger to the little finger* and *take away*, in order, *from the little finger to the index finger*.

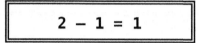

We do this as follows:

   that is, 1

For this problem, you should say, "One two minus one, that is 1".

$$4 - 2 = 2$$

We do this as follows:

  that is, 2

For this problem, you should say, "One two three four minus one two, that is 2".

Do Exercises 2.2 and 2.3, page 100

2.3 Basic addition with fingers F1–F4 and thumb F5

In the following, do not set 2, 3, 4 or 5 in one movement; *count* until you get 2, 3, 4 or 5. *Remember that you add, in order, from the index finger to the little finger.*

$$2 + 5 = 7$$

We do this as follows:

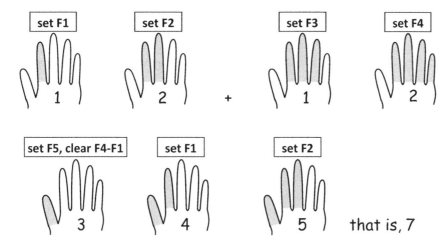

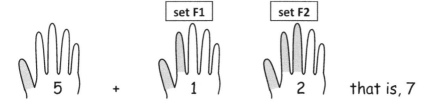

$$5 + 2 = 7$$

We do this as follows: Count from 1 until you get to 5 (first hand below). Then proceed by saying "plus"...

  Do Exercises 2.4 and 2.5, page 101

2.4 Basic subtraction with fingers F1–F4 and thumb F5

In the following, do not set a number in one movement; *count* until you get to the number. When minusing, take away one at a time. *Remember that you add, in order, from the index finger to the little finger and take away, in order, from the little finger to the index finger.*

$$7 - 2 = 5$$

We do this as follows:

- count from 1 to 7, saying each number as you set it (first hand below). Proceed by saying "minus"…

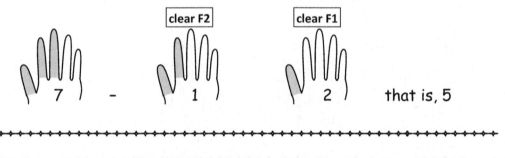

$$8 - 3 = 5$$

We do this as follows:

- count from 1 to 8, saying each number as you set it (first hand below). Proceed by saying "minus"…

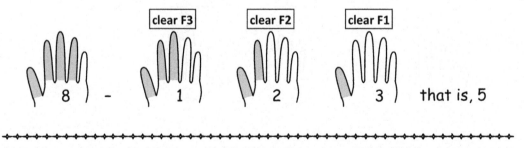

Do Exercises 2.6 and 2.7, pages 102-103

2.5 Basic addition/subtraction (F1-F5) without counting

So far, we have emphasized that addition and subtraction should be done by counting. This is important in understanding the concepts of adding and subtracting. However, to calculate quickly using your hands, you must be able to set or clear numbers in one movement. We show how in the following examples.

$$2 + 2 + 5 = 9$$

We do this as follows:

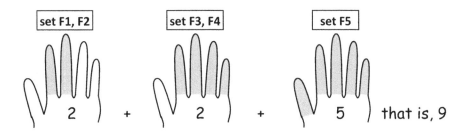

Note that to add 5, we simply set F5.

$$8 - 5 = 3$$

We do this as follows:

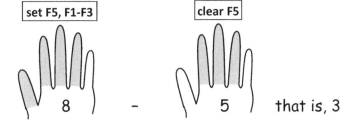

Note that to minus 5, we simply clear F5.

$$9 - 6 = 3$$

We do this as follows:

set F5, F1-F4	clear F5, F4

 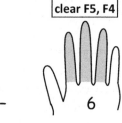

9   -   6   that is, 3

Note that to minus 6, we must clear F5, F4. We *always clear* fingers *from* the *little finger to* the index finger. In this example, **it would be *wrong* to clear F5, F1** since we would end up with F2, F3, F4 set

and this does not represent anything!

+++

$$3 + 6 - 5 = 4$$

We do this as follows:

set F1-F3	set F5, F4	clear F5

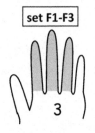 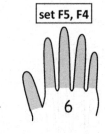 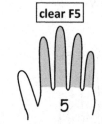

3   +   6   -   5   that is, 4

+++

Repeat Exercises 2.1-2.7, pages 100-103, without counting

2.6 Addition requiring a switch from 4 to 5

Drill: Practise counting from 1 to 5. Observe the 'switch' which takes place as you go from 4 to 5. You exchange F1–F4 for F5.

Goal: to get comfortable with the movement needed to go from 4 to 5

$$3 + 2 = 5$$

We do this as follows:

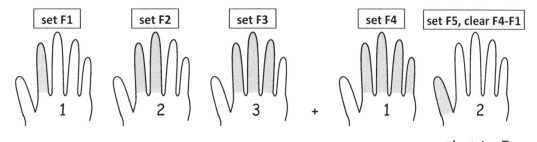

that is, 5

$$2 + 4 = 6$$

We do this as follows:

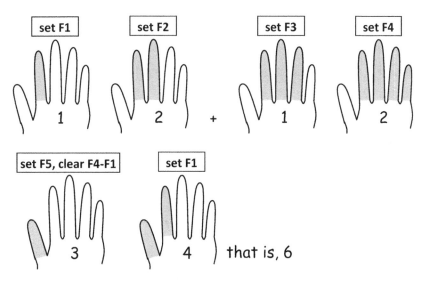

that is, 6

$$4 + 4 = 8$$

We do this as follows:

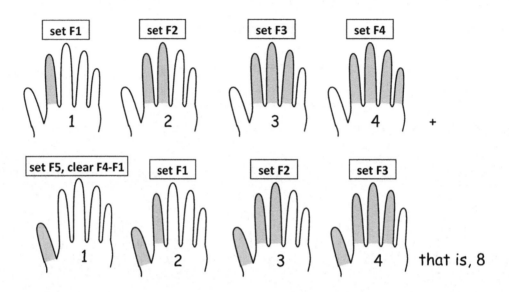

In this example, see if you can observe how the answer can be obtained in one step. We want

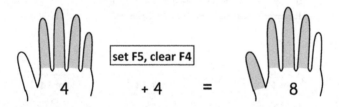

Note that to add 4, you set the thumb (+5) and clear F4 (−1) and **5 - 1** is **+4**.

Practise going from 4 to 8 in one step. Get a feel for dropping the thumb (F5) and lifting the little finger (F4).

Also practise adding 4 by going from 3 to 7, 2 to 6 and 1 to 5 in one step.

In chapter 5 (page 47), we will deal with this topic in more detail.

Do Exercises 2.8 and 2.9, pages 103-104

2.7 Subtraction requiring a switch from 5 to 4

Drill: Practise counting from 9 to 4. Observe the switch which takes place as you go from 5 to 4. You exchange F5 for F1–F4.

Goal: to get comfortable with the movement needed to go from 5 to 4

$$7 - 3 = 4$$

We do this as follows:

- count from 1 to 7, saying each number as you set it (first hand below). Proceed by saying "minus"…

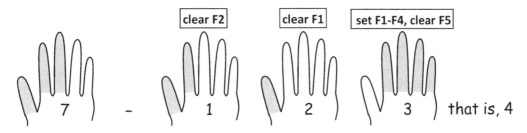

$$6 - 4 = 2$$

We do this as follows:

- count from 1 to 6, saying each number as you set it (first hand below). Proceed by saying "minus"…

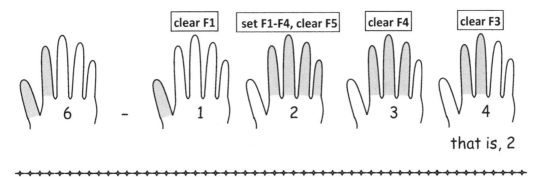

$$7 - 4 = 3$$

We do this as follows:

- count from 1 to 7, saying each number as you set it (first hand below). Proceed by saying "minus"...

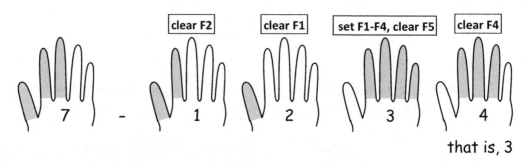

that is, 3

In this example, see if you can observe how the answer can be obtained in one step. We want

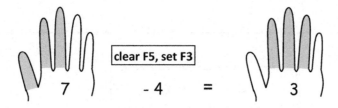

Notice that to minus 4, you clear the thumb (-5) and set F3 (+1) and **-5 + 1** is **-4**.

Practise going from 7 to 3 in one step. Get a feel for lifting the thumb (F5) and dropping the ring finger (F3).

Also practise minusing 4 by going from 8 to 4, 6 to 2 and 5 to 1 in one step.

In chapter 6 (page 53), we will deal with this topic in more detail.

> Do Exercises 2.10 and 2.11, pages 104-105

2.8 Revision with speed

Repeat Exercises 2.1-2.11 but, this time, do each problem as quickly as possible. For example, if the first number is 7, do not count to 7, just set F5, F1-F2. Also, if you have to add 4, just set F1-F4 and if that's not possible, add 5 and minus 1 in one step. Proceed as in the following examples.

$$3 + 4 = 7$$

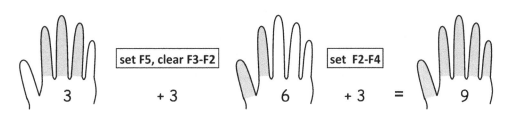

$$3 + 3 + 3 = 9$$

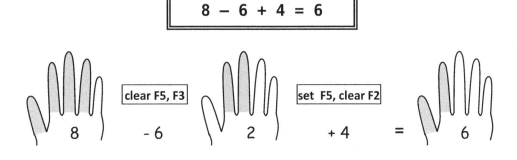

$$8 - 6 + 4 = 6$$

We will deal with this topic in more detail in chapters 5 and 6.

3
Basic 2-digit arithmetic

In this chapter, we introduce you to arithmetic using both hands. Recall that to store the number 37, say, we simply store 3 on the left hand and 7 on the right. The idea is to get you accustomed to using both hands.

The problems are simple in that when you do addition, there will be no "carry over" value from the right hand to the left. Also, when you do subtraction, there will be no need to "borrow" from the left hand. In other words, all the problems will be such that the calculation on one hand will not affect the calculation on the other. For example, to do 23 + 15, we simply add 2 and 1 on the left hand, and 3 and 5 on the right. The calculations are done separately on each hand.

In chapter 11 (page 77), we will show you how to do 2-digit problems which require "carry" and "borrow" from one hand to the other.

3.1 Add 2-digit and 1-digit numbers with no carry

In this section, we consider problems where we add a single-digit number to a 2-digit number and there is no carry from the right hand to the left hand. For example, consider

$$12 + 5 = 17$$

$$\begin{array}{r} 1\,2 \\ +\ 5 \end{array}$$

Store **12** by storing **1** on the left hand and **2** on the right

Add **5** to the right hand, giving

$$45 + 3 = 48$$

$$\begin{array}{r} 4\,5 \\ +\ 3 \end{array}$$

Store **45** by storing **4** on the left hand and **5** on the right

Add **3** to the right hand, giving

$$72 + 4 = 76$$

$$\begin{array}{r} 7\,2 \\ +\ 4 \end{array}$$

Store **72** by storing **7** on the left hand and **2** on the right

Add **4** to the right hand, giving

Do Exercise 3.1, page 106

3.2 Add 2-digit numbers with no carry

We now consider problems where we add a 2-digit number to a 2-digit number and there is no carry from the right hand to the left hand. For example, consider

$$13 + 21 = 34$$

$$\begin{array}{r} 1\,3 \\ +2\,1 \end{array}$$

Store **13** by storing **1** on the left hand and **3** on the right

Add **2** to the left hand, giving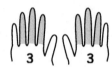

Add **1** to the right hand, giving

In DigitalMath (and mental arithmetic, in general), it is usually faster to work from left to right.

$$63 + 24 = 87$$

$$\begin{array}{r} 6\,3 \\ +2\,4 \end{array}$$

Store **63** by storing **6** on the left hand and **3** on the right

Add **2** to the left hand, giving

Add **4** to the right hand, giving

 Do Exercise 3.2, page 106

We will see how to add 2-digit number 'with carry' in chapter 11 (page 77).

3.3 Subtract 1-digit from 2-digit with no borrow

We consider problems where we subtract a 1-digit number from a 2-digit number and we do not need to borrow. For example, consider

$$78 - 4 = 74$$

$$\begin{array}{r} 7\,8 \\ -\ \ 4 \end{array}$$

Store **78** by storing **7** on the left hand and **8** on the right

Subtract **4** from the right hand, giving

$$49 - 6 = 43$$

$$\begin{array}{r} 4\,9 \\ -\ \ 6 \end{array}$$

Store **49** by storing **4** on the left hand and **9** on the right

Subtract **6** from the right hand, giving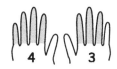

Do Exercise 3.3, page 107

3.4 Subtract 2-digit numbers with no borrow

We consider problems where we subtract a 2-digit number from another 2-digit number and we do not need to borrow. For example, consider

$$78 - 26 = 52$$

$$\begin{array}{r} 7\,8 \\ -2\,6 \end{array}$$

Store **78** by storing **7** on the left hand and **8** on the right

Subtract **2** from the left hand, giving

Subtract **6** from the right hand, giving

$$85 - 53 = 32$$

$$\begin{array}{r} 8\,5 \\ -5\,3 \end{array}$$

Store **85** by storing **8** on the left hand and **5** on the right

Subtract **5** from the left hand, giving

Subtract **3** from the right hand, giving

Do Exercises 3.4 and 3.5, pages 108-109

We will see how to subtract 2-digit number 'with borrow' in chapter 11.

4
General single-digit addition/subtraction

In this chapter, we consider problems in which the numbers involved could be bigger than 9. Since you cannot store more than 9 on the right hand, these problems would require you to use both hands. For example, if you wish to add 6 to 8, both hands must come into play since the answer, 14, is too big for the right hand only.

4.1 General single-digit addition

We now take a look at problems where the answer from adding single-digit numbers can be bigger than 9.

Drill: Practise counting from 5 to 10. Observe the 'switch' which takes place as you go from 9 to 10. You exchange the entire right hand for F10.

Continue to practise by counting from 15 to 20, 25 to 30, 35 to 40, 45 to 50, 55 to 60, 65 to 70, 75 to 80 and 85 to 90.

Goal: to get comfortable with the movement needed to go from units to tens

$$8 + 4 = 12$$

We do this as follows:

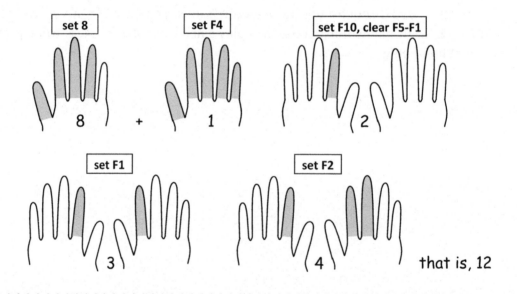

 that is, 12

The key move here is going from 9 to 10. If we were writing 10, we would write "1" followed by "0". In the same way, we "write 1" on our left hand (set F10) and "0" on our right (clear F5-F1). It might be helpful to exaggerate the lifting of the entire right hand from the table as you go from 9 to 10.

7 + 6 = 13

We do this as follows:

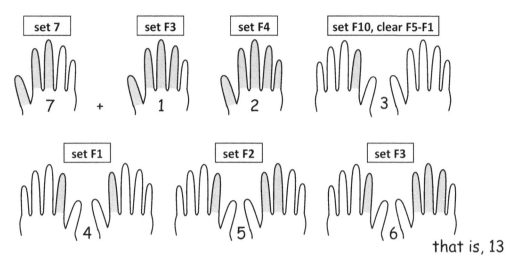

that is, 13

---

5 + 5 = 10

We do this as follows:

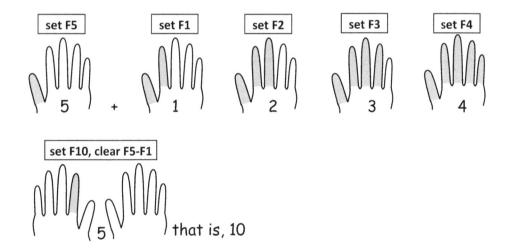

that is, 10

Note that **to add 5**, we **add 1 to the left hand** (which has the value 10) **and minus 5** (clear F5) **from the right hand**. In effect, we do **+10 − 5** which is **+5**. After you fully understand how to add **5** by counting, you can use this movement to speed things up, as in the following examples.

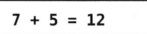

$$7 + 5 = 12$$

set F5, F1, F2   set F10, clear F5

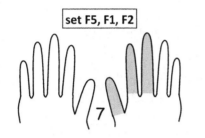   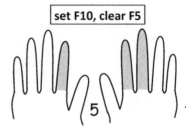

7   +   5   ) that is, 12

$$18 + 5 = 23$$

set F10, F5, F1-F3   set F20, clear F5

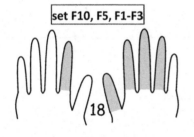   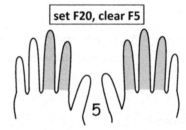

18   +   5   ) that is, 23

Here, we add 10 (which is the same as adding 1 to the left hand) and minus 5 from the right hand.

Do Exercises 4.1 and 4.2, page 110

Do Exercises 4.3 and 4.4, pages 111-113

4.2 General single-digit subtraction

We now take a look at subtraction problems which involve both hands.

Drill: Practise counting from 15 down to 0. Observe the 'switch' which takes place as you go from 10 to 9. You exchange F10 for the entire right hand (9). As an added bonus, you also get to revise the switch from 5 to 4.

Goal: to get comfortable with the movement needed to go from tens to units

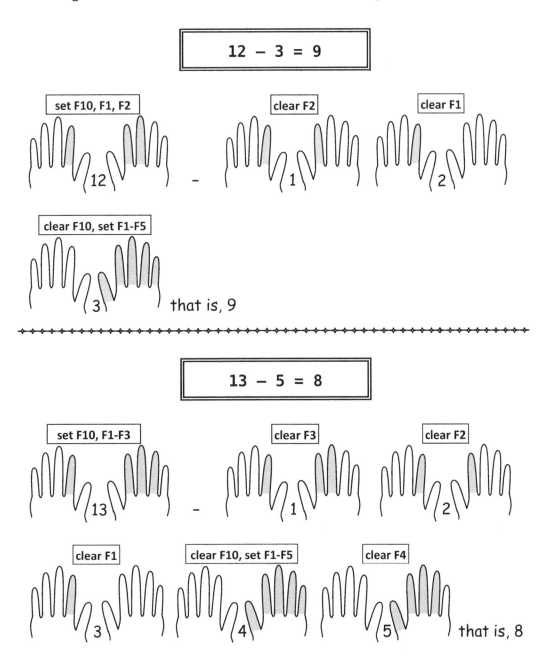

that is, 9

that is, 8

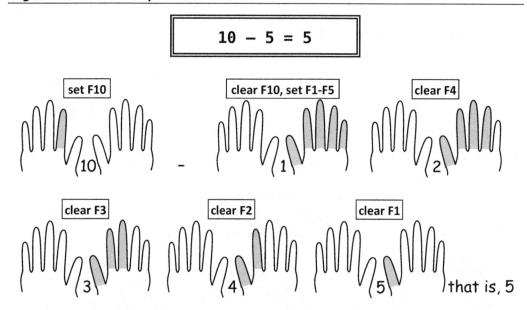

Note that **to minus 5, you simply do a switch of F10 with F5** (clear F10, set F5). In effect, this is −10 + 5 which is −5. After you fully understand how to minus 5 by counting, you can use this movement to speed things up.

In the next example, we need to count to 9 and we count starting with 5.

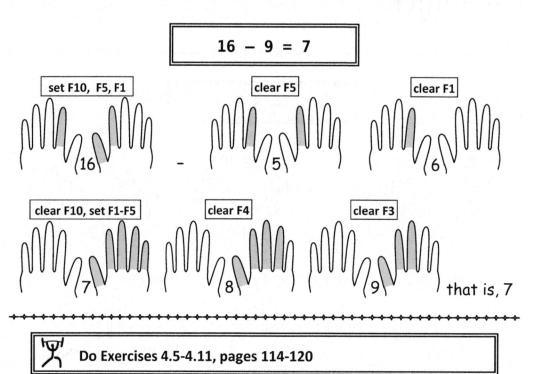

Do Exercises 4.5-4.11, pages 114-120

5
Faster way to add on right hand

So far, we have done addition by "counting". For instance, if we want to add 3, we count 1, 2, 3. This is fine and it works for all problems. But, sometimes, we may need to go a little faster. Previously, we have hinted at some of these techniques. For example, to add 4 to 3, you can simply add 5 (set F5) and minus 1 (clear F3), instead of counting 1, 2, 3, 4. In this chapter, we discuss, in more detail, some of the techniques you can use to speed up your calculations.

The general idea is that if you want to add a number and you cannot do so directly, you can add a bigger number (like 5 or 10) and subtract the right amount to get the number you want. For example, if you have 3 on your right hand (F1, F2, F3) and you want to **add 4**, you can do so **by adding 5** (set F5) **and subtracting 1** (clear F3), like this:

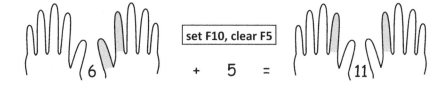

Similarly, suppose you have 6 on your right hand (F5, F1) and you want to add 5. Since you do not have 5 available on that hand, you can add 1 to the left hand (which has the value 10) and subtract 5 from the right hand (clear F5), like this:

We now look at faster ways to add numbers from 1 to 5.

5.1 Fast way to add 4 on right hand: + 4 = + 5 – 1

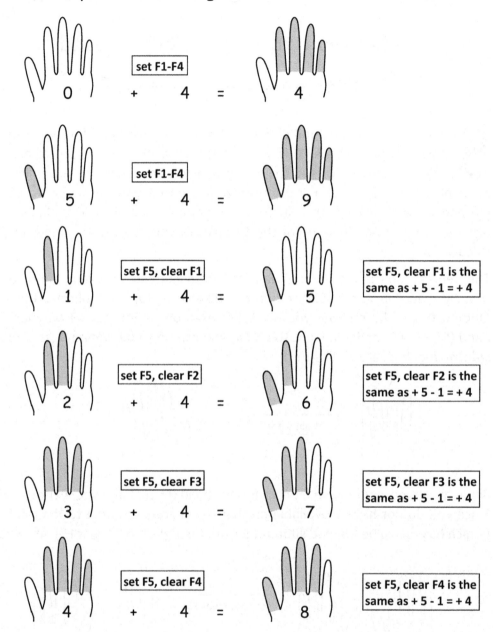

0	set F1-F4 + 4 =	4
5	set F1-F4 + 4 =	9
1	set F5, clear F1 + 4 =	5 — set F5, clear F1 is the same as + 5 - 1 = + 4
2	set F5, clear F2 + 4 =	6 — set F5, clear F2 is the same as + 5 - 1 = + 4
3	set F5, clear F3 + 4 =	7 — set F5, clear F3 is the same as + 5 - 1 = + 4
4	set F5, clear F4 + 4 =	8 — set F5, clear F4 is the same as + 5 - 1 = + 4

Practice: 5 + 4, 1 + 4, 2 + 4, 3 + 4, 4 + 4 : get the feel for dropping the thumb and lifting a finger.

Do Exercise 5.1, page 121

5.2 Fast way to add 3 on right hand: + 3 = + 5 – 2

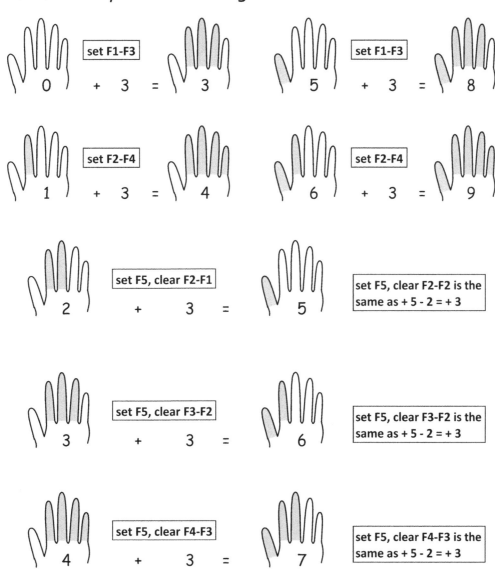

+ +

Practice: 1 + 3, 5 + 3, 6 + 3, 2 + 3, 3 + 3, 4 + 3 : get the feel for dropping the thumb and lifting two fingers.

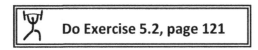 **Do Exercise 5.2, page 121**

5.3 Fast way to add 2 on right hand: + 2 = + 5 – 3

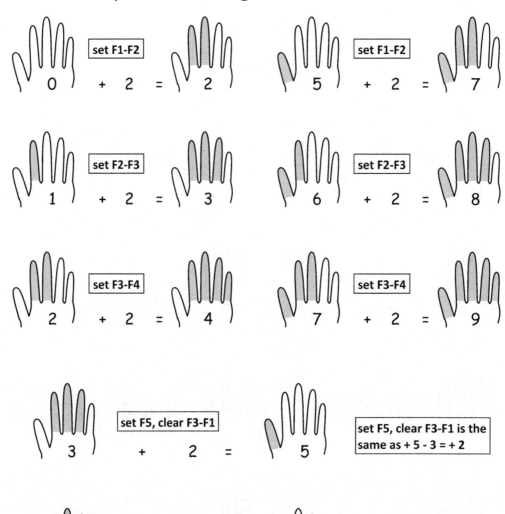

Practice: 1 + 2, 2 + 2, 5 + 2, 6 + 2, 7 + 2, 3 + 2, 4 + 2 : get the feel for dropping the thumb and lifting three fingers.

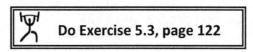

Do Exercise 5.3, page 122

5.4 Ways to add 1: + 1 = + 5 – 4 = + 10 – 9 = – 9 + 10

If you learnt your counting well, you will already know how to add 1 in different situations. There is no new way to add 1; we will just show you another way of looking at it.

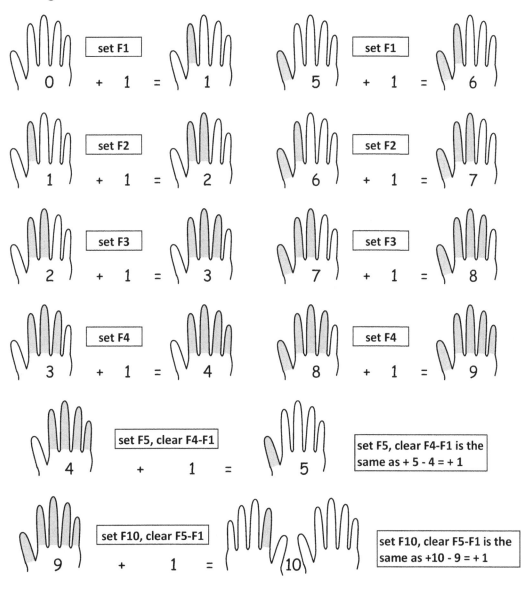

set F1
0 + 1 = 1

set F1
5 + 1 = 6

set F2
1 + 1 = 2

set F2
6 + 1 = 7

set F3
2 + 1 = 3

set F3
7 + 1 = 8

set F4
3 + 1 = 4

set F4
8 + 1 = 9

set F5, clear F4-F1
4 + 1 = 5

set F5, clear F4-F1 is the same as + 5 - 4 = + 1

set F10, clear F5-F1
9 + 1 = 10

set F10, clear F5-F1 is the same as +10 - 9 = + 1

Practice: 1 + 1, 2 + 1, 3 + 1, 5 + 1, 6 + 1, 7 + 1, 8 + 1, 4 + 1, 9 + 1

Do Exercise 5.4, page 123

5.5 Fast way to add 5: + 5 = + 10 – 5 = – 5 + 10

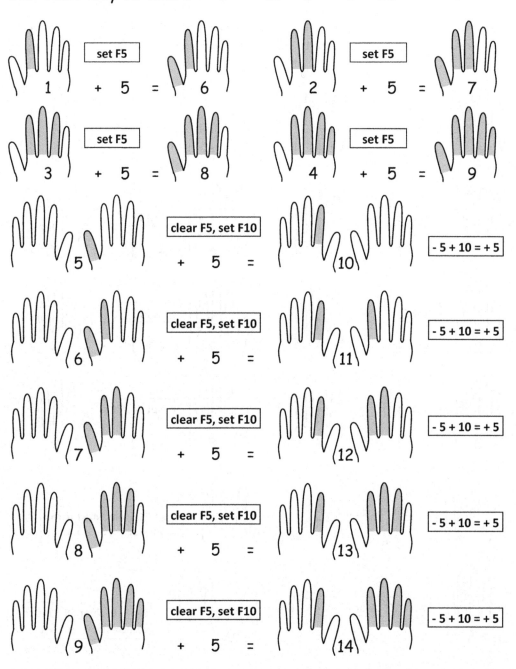

Practice: 1 + 5, 2 + 5, 3 + 5, 4 + 5, 5 + 5, 6 + 5, 7 + 5, 8 + 5, 9 + 5

Do Exercises 5.5 and 5.6, pages 123-124

6
Faster way to subtract from right hand

So far, we have done subtraction by "counting". For instance, if we want to minus 3, we count 1, 2, 3, as we remove fingers. This is fine and it works for all problems. But, sometimes, we may need to go a little faster. In this chapter, we discuss some of the techniques you can use to speed up your calculations.

The general idea is that if you want to subtract a number and you cannot do so directly, you can subtract a bigger number (like 5 or 10) and add the right amount to get the number you want. For example, if you have 7 on your right hand (F5, F1, F2) and you want to *minus 4*, you can do so *by subtracting 5* (clear F5) *and adding 1* (set F3), like this:

Similarly, suppose you have 5 on your right hand (F5) and you want to minus 1. You can simply count backwards from 5 to 4 but you can think of it like this: since you do not have a single finger to clear, you can minus 5 (clear F5) and add 4 (set F1-F4), like this:

We now look at faster ways to subtract 4, 3, 2 and 1 from the right hand.

6.1 Fast way to minus 4 from right hand: - 4 = - 5 + 1

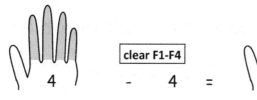

4 | clear F1-F4 | -   4   =   0

9 | clear F1-F4 | -   4   =   5

 5 | clear F5, set F1 | -   4   =   1 | clear F5, set F1 is the same as - 5 + 1 = - 4 |

6 | clear F5, set F2 | -   4   =   2 | clear F5, set F2 is the same as - 5 + 1 = - 4 |

 7 | clear F5, set F3 | -   4   =   3 | clear F5, set F3 is the same as - 5 + 1 = - 4 |

8 | clear F5, set F4 | -   4   =   4 | clear F5, set F4 is the same as - 5 + 1 = - 4 |

Practice: 4 – 4, 9 – 4, 5 – 4, 6 – 4, 7 – 4, 8 – 4

Do Exercise 6.1, page 125

6.2 Fast way to minus 3 from right hand: - 3 = - 5 + 2

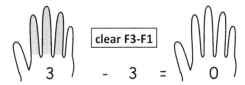

3 clear F3-F1 $- 3 =$ 0 8 clear F3-F1 $- 3 =$ 5

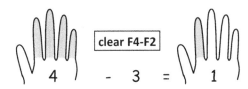

4 clear F4-F2 $- 3 =$ 1 9 clear F4-F2 $- 3 =$ 6

5 clear F5, set F1-F2 $- 3 =$ 2 clear F5, set F1-F2 is the same as - 5 + 2 = - 3

6 clear F5, set F2-F3 $- 3 =$ 3 clear F5, set F2-F3 is the same as - 5 + 2 = - 3

7 clear F5, set F3-F4 $- 3 =$ 4 clear F5, set F3-F4 is the same as - 5 + 2 = - 3

Practice: $3 - 3, 4 - 3, 8 - 3, 9 - 3, 5 - 3, 6 - 3, 7 - 3$

Do Exercise 6.2, page 125

6.3 Fast way to minus 2 from right hand: - 2 = - 5 + 3

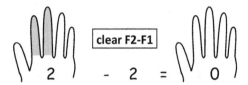

2   clear F2-F1   - 2 = 0   7   clear F2-F1   - 2 = 5

3   clear F3-F2   - 2 = 1   8   clear F3-F2   - 2 = 6

4   clear F4-F3   - 2 = 2   9   clear F4-F3   - 2 = 7

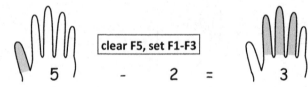

5   clear F5, set F1-F3   - 2 = 3   clear F5, set F1-F3 is the same as - 5 + 3 = - 2

6   clear F5, set F2-F4   - 2 = 4   clear F5, set F2-F4 is the same as - 5 + 3 = - 2

+++

Practice: 2 – 2, 3 – 2, 4 – 2, 7 – 2, 8 – 2, 9 – 2, 5 – 2, 6 – 2

> **Do Exercise 6.3, page 126**

6.4 Ways to minus 1: -1 = - 5 + 4 = - 10 + 9

If you learnt how to count backwards, you will already know how to subtract 1 in different situations. There is no new way to subtract 1; we will just show you another way of looking at it.

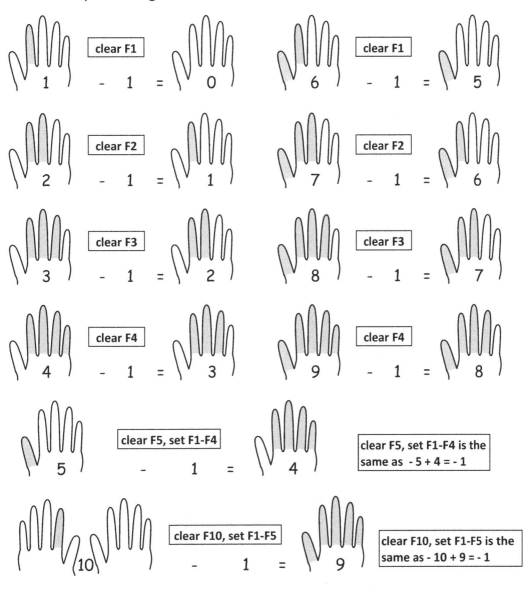

Practice: $1 - 1$, $2 - 1$, $3 - 1$, $4 - 1$, $6 - 1$, $7 - 1$, $8 - 1$, $9 - 1$, $5 - 1$, $10 - 1$

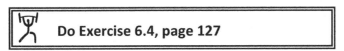 **Do Exercise 6.4, page 127**

7
Faster way to add 9, 8, 7, 6

So far, we have shown you how to add any single-digit number by counting. This will work all the time but it is a bit slow when you need to calculate quickly. Consider how we might add 7.

If the right hand has values 0, 1 or 2, we can easily add 7 to this hand. But what if the hand has a value of 3 or more? Adding 7 will give us an answer that is more than 9—and this is too much to store on the right hand, so we must use our left.

What we do is *add 1 to the left hand* (which, remember, has a value of 10) and *minus 3 from the right hand*. This will work since **10 − 3 = 7**. In the following example, we want to add 7 to 8. To do so, we add 1 to the left hand and minus 3 from the right, giving us the answer **15**.

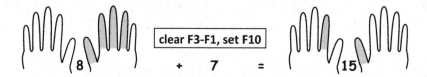

clear F3-F1, set F10

8   +   7   =   15

In DigitalMath, it is usual to write **+ 7 = - 3 + 10**. The reason is that when we have to add 7, we must first look at our right hand to see if 7 can be added to that hand. If the hand contains 2 or less, we can. But if it contains 3 or more, we can't. Since our *focus* is on that hand, it is faster to minus 3 *now*, rather than change focus to the left hand to minus 10 and *then* minus 3 from the right hand.

While this is useful as you learn the technique, in the end, it does not matter too much since you should reach the stage where you can **add 10 and minus 3 at the same time**. For the moment, if you prefer to add 10 first, feel free to do so.

7.1 Fast way to add 9: + 9 = + 10 – 1 = – 1 + 10

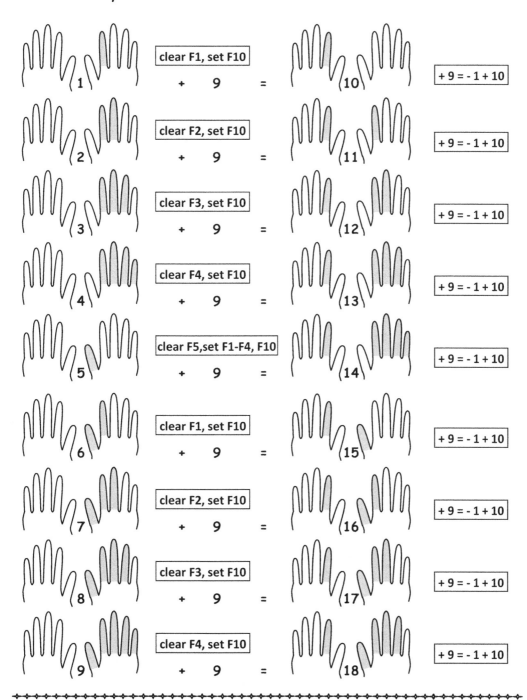

1	clear F1, set F10 + 9 =	10	+ 9 = – 1 + 10
2	clear F2, set F10 + 9 =	11	+ 9 = – 1 + 10
3	clear F3, set F10 + 9 =	12	+ 9 = – 1 + 10
4	clear F4, set F10 + 9 =	13	+ 9 = – 1 + 10
5	clear F5,set F1-F4, F10 + 9 =	14	+ 9 = – 1 + 10
6	clear F1, set F10 + 9 =	15	+ 9 = – 1 + 10
7	clear F2, set F10 + 9 =	16	+ 9 = – 1 + 10
8	clear F3, set F10 + 9 =	17	+ 9 = – 1 + 10
9	clear F4, set F10 + 9 =	18	+ 9 = – 1 + 10

Practice: 1 + 9, 2 + 9, 3 + 9, 4 + 9, 6 + 9, 7 + 9, 8 + 9, 9 + 9, 5 + 9

Do Exercise 7.1, page 127

7.2 Fast way to add 8: + 8 = + 10 − 2 = − 2 + 10

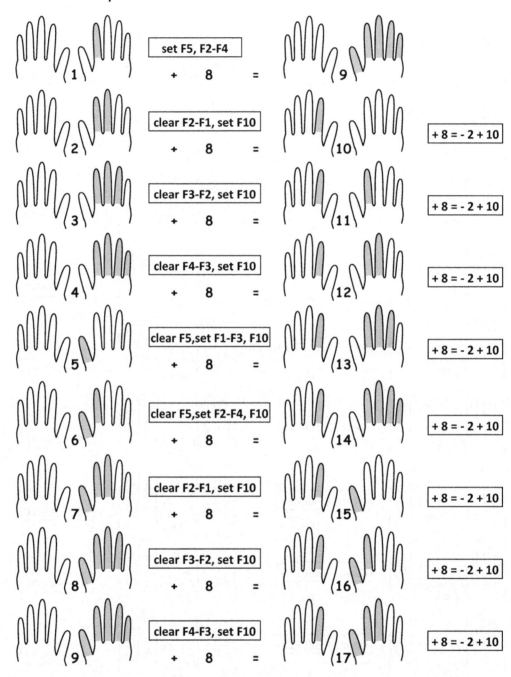

set F5, F2-F4	
+ 8 =	
clear F2-F1, set F10	+ 8 = − 2 + 10
+ 8 =	
clear F3-F2, set F10	+ 8 = − 2 + 10
+ 8 =	
clear F4-F3, set F10	+ 8 = − 2 + 10
+ 8 =	
clear F5,set F1-F3, F10	+ 8 = − 2 + 10
+ 8 =	
clear F5,set F2-F4, F10	+ 8 = − 2 + 10
+ 8 =	
clear F2-F1, set F10	+ 8 = − 2 + 10
+ 8 =	
clear F3-F2, set F10	+ 8 = − 2 + 10
+ 8 =	
clear F4-F3, set F10	+ 8 = − 2 + 10
+ 8 =	

Practice: 1 + 8, 2 + 8, 3 + 8, 4 + 8, 7 + 8, 8 + 8, 9 + 8, 5 + 8, 6 + 8

Do Exercise 7.2, page 128

7.3 Fast way to add 7: + 7 = + 10 – 3 = - 3 + 10

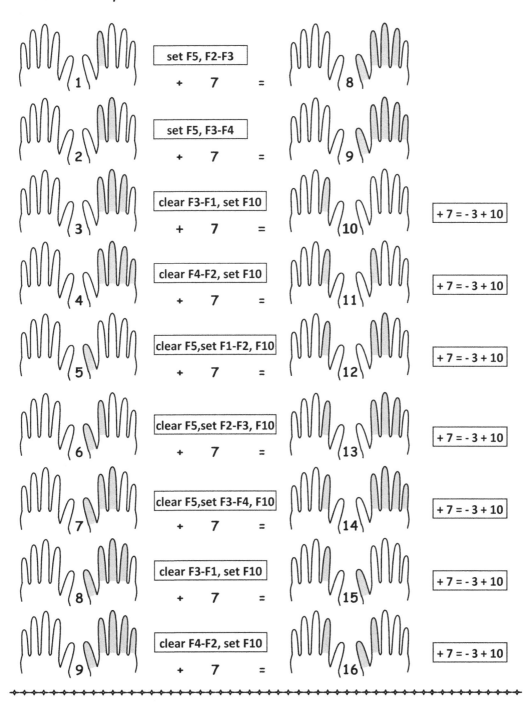

set F5, F2-F3

+   7   =

set F5, F3-F4

+   7   =

clear F3-F1, set F10

+   7   =

clear F4-F2, set F10

+   7   =

clear F5, set F1-F2, F10

+   7   =

clear F5, set F2-F3, F10

+   7   =

clear F5, set F3-F4, F10

+   7   =

clear F3-F1, set F10

+   7   =

clear F4-F2, set F10

+   7   =

+ 7 = - 3 + 10

+ 7 = - 3 + 10

+ 7 = - 3 + 10

+ 7 = - 3 + 10

+ 7 = - 3 + 10

+ 7 = - 3 + 10

+ 7 = - 3 + 10

Practice: 1 + 7, 2 + 7, 3 + 7, 4 + 7, 8 + 7, 9 + 7, 5 + 7, 6 + 7, 7 + 7

Do Exercise 7.3, page 128

7.4 Fast way to add 6: + 6 = + 10 − 4 = − 4 + 10

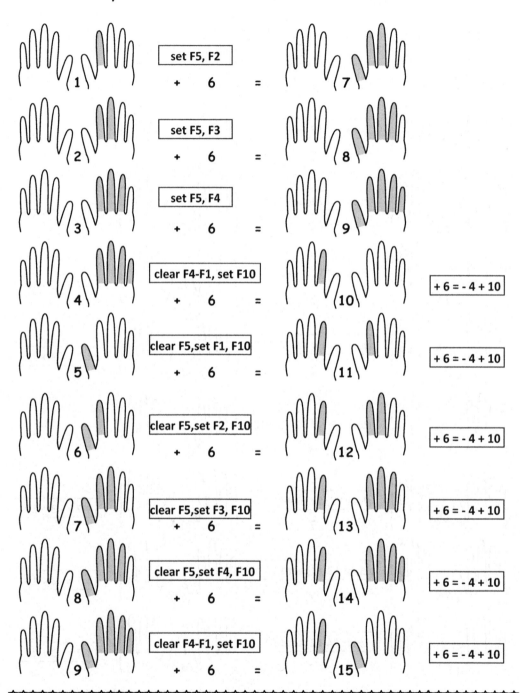

set F5, F2	+ 6 =	(7)
set F5, F3	+ 6 =	(8)
set F5, F4	+ 6 =	(9)
clear F4-F1, set F10	+ 6 =	(10) + 6 = − 4 + 10
clear F5,set F1, F10	+ 6 =	(11) + 6 = − 4 + 10
clear F5,set F2, F10	+ 6 =	(12) + 6 = − 4 + 10
clear F5,set F3, F10	+ 6 =	(13) + 6 = − 4 + 10
clear F5,set F4, F10	+ 6 =	(14) + 6 = − 4 + 10
clear F4-F1, set F10	+ 6 =	(15) + 6 = − 4 + 10

Practice: 1 + 6, 2 + 6, 3 + 6, 4 + 6, 9 + 6, 5 + 6, 6 + 6, 7 + 6, 8 + 6

Do Exercises 7.4, 7.5, 7.6, pages 129-131

$$\boxed{\begin{array}{c} \textbf{8} \\ \textbf{Add 4, 3, 2 with carry to left hand} \end{array}}$$

I n this chapter, we show you how to add 4, 3 and 2 when the resulting answer is too big to store on the right hand. We will use the following facts:

$$+ 4 = + 10 - 6 = - 6 + 10$$
$$+ 3 = + 10 - 7 = - 7 + 10$$
$$+ 2 = + 10 - 8 = - 8 + 10$$

Our preference is to minus from the right hand first and then add 10 to the left hand. But some people prefer to do it the other way around so we won't make a rule about it. We note, however, that with a little practice, you will find yourself performing both operations *at the same time* so the question of which to do first does not arise.

The general idea is that if you want to add a number (4, say) and you do not have enough on the right hand, you can add 10 to the left hand (by adding 1) and take away 6 from the right hand. For example, we show how to add 4 to 8:

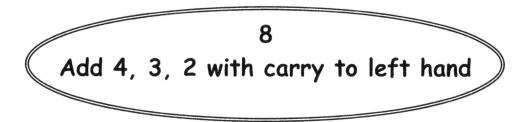

clear F5, F3 (- 6)
set F10 (+ 10)

8   +   4   =   12

We now show you how to add 4, 3 and 2 when there's a carry to the left hand.

8.1 Add 4 with carry to left hand: + 4 = - 6 + 10

When the right hand has a value of 6 or more, 4 cannot be added to this hand directly. **To add 4**, we must **add 1 to the left hand** (which, remember, has a value of 10) **and subtract 6 from the right hand**. We have **10 – 6 = 4**.

However, note that we write this as **+ 4 = - 6 + 10**. The reason is that when we have to add 4, we must first look at our right hand to see if 4 can be added to that hand. If the hand contains 5 or less, we can. But if it contains 6 or more, we can't. Since our *focus* is on that hand, it is faster to minus 6 *now*, rather than change focus to the left hand to minus 10 and *then* minus 6 from the right hand.

While this is useful as you learn the technique, in the end, you should reach the stage where you can **add 10 and minus 6 at the same time**.

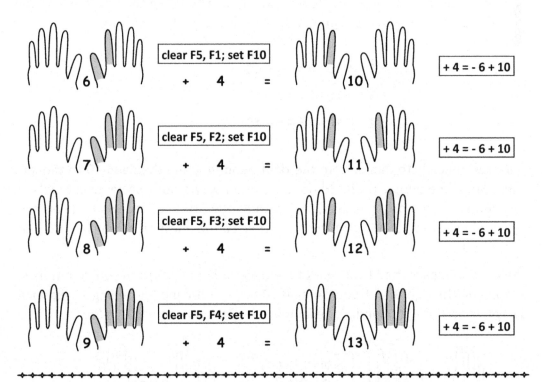

clear F5, F1; set F10	+ 4 =	+ 4 = - 6 + 10 (10)
clear F5, F2; set F10	+ 4 =	+ 4 = - 6 + 10 (11)
clear F5, F3; set F10	+ 4 =	+ 4 = - 6 + 10 (12)
clear F5, F4; set F10	+ 4 =	+ 4 = - 6 + 10 (13)

Practice: 1 + 4, 2 + 4, 3 + 4, 4 + 4, 5 + 4, 6 + 4, 7 + 4, 8 + 4, 9 + 4

Do Exercise 8.1, page 132

8.2 Add 3 with carry to left hand: + 3 = - 7 + 10

When the right hand has a value of 7 or more, 3 cannot be added to this hand directly. **To add 3**, we must **add 1 to the left hand** (which, remember, has a value of 10) **and subtract 7 from the right hand**. We have **10 – 7 = 3**.

However, note that we write this as **+ 3 = - 7 + 10**. The reason is that when we have to add 3, we must first look at our right hand to see if 3 can be added to that hand. If the hand contains 6 or less, we can. But if it contains 7 or more, we can't. Since our *focus* is on that hand, it is faster to minus 7 *now*, rather than change focus to the left hand to minus 10 and *then* minus 7 from the right hand.

While this is useful as you learn the technique, in the end, you should reach the stage where you can **add 10 and minus 7 at the same time**.

7	clear F5, F1-F2; set F10 + 3 =	10 + 3 = - 7 + 10
8	clear F5, F2-F3; set F10 + 3 =	11 + 3 = - 7 + 10
9	clear F5, F3-F4; set F10 + 3 =	12 + 3 = - 7 + 10

Practice: 1 + 3, 2 + 3, 3 + 3, 4 + 3, 5 + 3, 6 + 3, 7 + 3, 8 + 3, 9 + 3

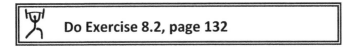

Do Exercise 8.2, page 132

8.3 Add 2 with carry to left hand: + 2 = - 8 + 10

When the right hand has a value of 8 or 9, 2 cannot be added to this hand directly. **To add 2**, we must **add 1 to the left hand** (which, remember, has a value of 10) **and subtract 8 from the right hand**. We have **10 – 8 = 2**.

However, note that we write this as **+ 2 = - 8 + 10**. The reason is that when we have to add 2, we must first look at our right hand to see if 2 can be added to that hand. If the hand contains 7 or less, we can. But if it contains 8 or 9, we can't. Since our *focus* is on that hand, it is faster to minus 8 *now*, rather than change focus to the left hand to minus 10 and *then* minus 8 from the right hand.

While this is useful as you learn the technique, in the end, you should reach the stage where you can **add 10 and minus 8 at the same time**.

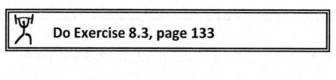

Practice: 1 + 2, 2 + 2, 3 + 2, 4 + 2, 5 + 2, 6 + 2, 7 + 2, 8 + 2, 9 + 2

Do Exercise 8.3, page 133

Do Exercise 8.4, page 133

9
Faster way to minus 5, 9, 8, 7, 6

I n this chapter, we show you how to minus 5, 9, 8, 7 and 6 when the right hand value is too small. For example, if we want to minus 8 from 12, the right hand contains only 2 and 8 cannot be subtracted from 2. In this case, we will minus 10 from the left hand and add 2 to the right hand, like this:

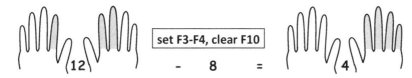

We will use the following facts:

$$- 5 = - 10 + 5 = + 5 - 10$$
$$- 9 = - 10 + 1 = + 1 - 10$$
$$- 8 = - 10 + 2 = + 2 - 10$$
$$- 7 = - 10 + 3 = + 3 - 10$$
$$- 6 = - 10 + 4 = + 4 - 10$$

Our preference is to add to the right hand first and then minus from the left hand. But some people prefer to do it the other way around so we won't make a rule about it. We note, however, that with a little practice, you should find yourself performing both operations *at the same time* so the question of which to do first does not arise.

9.1 Fast way to minus 5: - 5 = - 10 + 5 = + 5 - 10

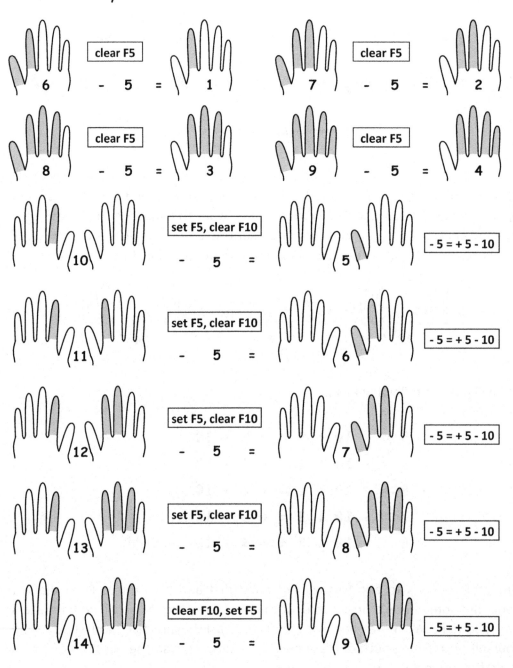

Practice: 5 – 5, 6 – 5, 7 – 5, 8 – 5, 9 – 5, 10 – 5, 11 – 5, 12 – 5, 13 – 5, 14 – 5

Do Exercise 9.1, page 134

9.2 Fast way to minus 9: - 9 = - 10 + 1 = + 1 - 10

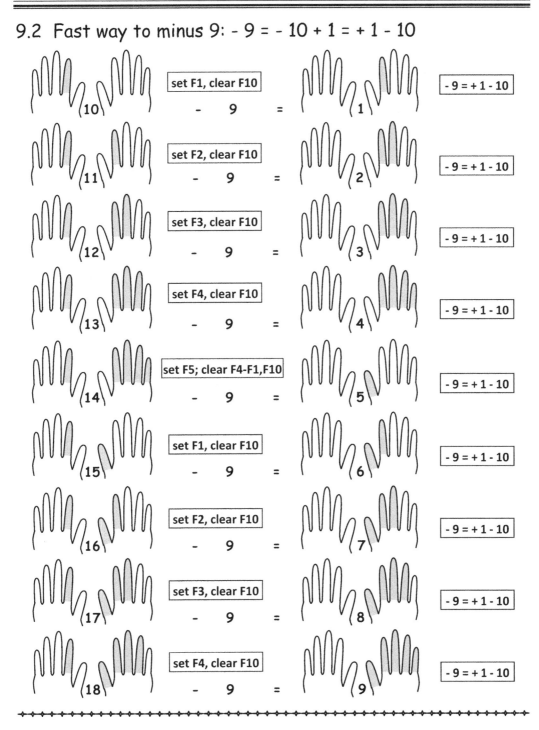

Practice: 10 - 9, 11 - 9, 12 - 9, 13 - 9, 14 - 9, 15 - 9, 16 - 9, 17 - 9, 18 - 9, 19 - 9

Do Exercise 9.2, page 135

9.3 Fast way to minus 8: - 8 = - 10 + 2 = + 2 - 10

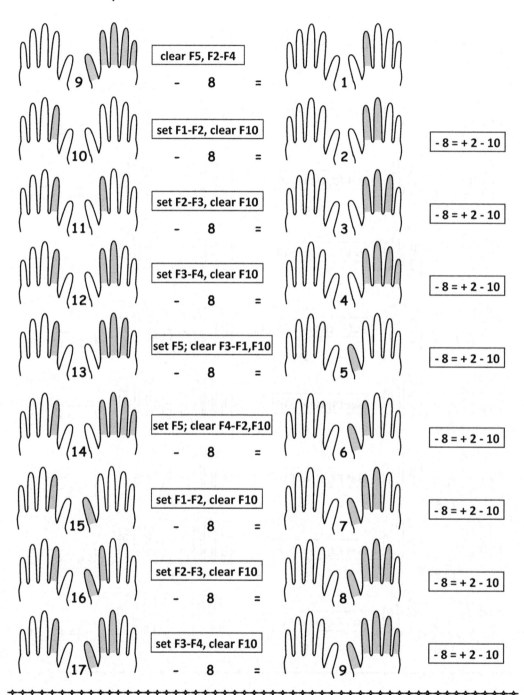

clear F5, F2-F4	
9 - 8 = 1	
set F1-F2, clear F10	
10 - 8 = 2	- 8 = + 2 - 10
set F2-F3, clear F10	
11 - 8 = 3	- 8 = + 2 - 10
set F3-F4, clear F10	
12 - 8 = 4	- 8 = + 2 - 10
set F5; clear F3-F1,F10	
13 - 8 = 5	- 8 = + 2 - 10
set F5; clear F4-F2,F10	
14 - 8 = 6	- 8 = + 2 - 10
set F1-F2, clear F10	
15 - 8 = 7	- 8 = + 2 - 10
set F2-F3, clear F10	
16 - 8 = 8	- 8 = + 2 - 10
set F3-F4, clear F10	
17 - 8 = 9	- 8 = + 2 - 10

Practice: 9 - 8, 10 - 8, 11 - 8, 12 - 8, 13 - 8, 14 - 8, 15 - 8, 16 - 8, 17 - 8, 18 − 8

Do Exercise 9.3, page 135

9.4 Fast way to minus 7: - 7 = - 10 + 3 = + 3 - 10

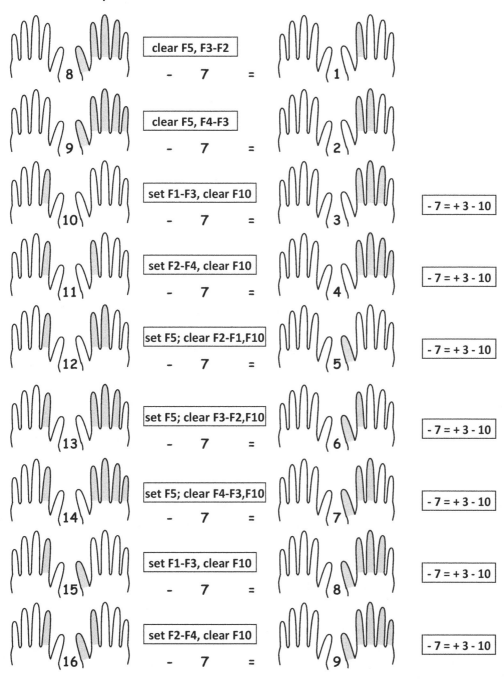

8	clear F5, F3-F2 - 7 =	1
9	clear F5, F4-F3 - 7 =	2
10	set F1-F3, clear F10 - 7 =	3 - 7 = + 3 - 10
11	set F2-F4, clear F10 - 7 =	4 - 7 = + 3 - 10
12	set F5; clear F2-F1,F10 - 7 =	5 - 7 = + 3 - 10
13	set F5; clear F3-F2,F10 - 7 =	6 - 7 = + 3 - 10
14	set F5; clear F4-F3,F10 - 7 =	7 - 7 = + 3 - 10
15	set F1-F3, clear F10 - 7 =	8 - 7 = + 3 - 10
16	set F2-F4, clear F10 - 7 =	9 - 7 = + 3 - 10

Practice: 8 - 7, 9 - 7, 10 - 7, 11 - 7, 12 - 7, 13 - 7, 14 - 7, 15 - 7, 16 - 7, 17 - 7

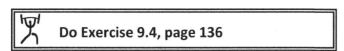

Do Exercise 9.4, page 136

9.5 Fast way to minus 6: - 6 = - 10 + 4 = + 4 - 10

7	clear F5, F2 - 6 =	1	
8	clear F5, F3 - 6 =	2	
9	clear F5, F4 - 6 =	3	
10	set F1-F4, clear F10 - 6 =	4	- 6 = + 4 - 10
11	set F5; clear F1, F10 - 6 =	5	- 6 = + 4 - 10
12	set F5; clear F2, F10 - 6 =	6	- 6 = + 4 - 10
13	set F5; clear F3, F10 - 6 =	7	- 6 = + 4 - 10
14	set F5; clear F4, F10 - 6 =	8	- 6 = + 4 - 10
15	set F1-F4, clear F10 - 6 =	9	- 6 = + 4 - 10

Practice: 7 - 6, 8 - 6, 9 - 6, 10 - 6, 11 - 6, 12 - 6, 13 - 6, 14 - 6, 15 - 6, 16 - 6

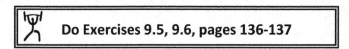

Do Exercises 9.5, 9.6, pages 136-137

10
Faster way to minus 4, 3, 2

When the right hand contains 4 or more, it is possible to subtract 4 from this hand. Similarly, when it contains 3 or more, we can subtract 3 and when it contains 2 or more, we can subtract 2. But what do we do when we want to subtract 4, say, and the right hand contains less than 4?

We do what we are accustomed to doing—we "borrow" 10 from the left hand. But we do things a bit differently from the way we were probably taught in school. If we want **to minus 4**, we **minus 10 from the left hand and add 6 to the right hand**. We have **-10 + 6 = -4**. For example, if we need to do **21 − 4**, it will be done as follows: **add 6 to the right hand** (set F5, F2) and **minus 1** (value 10) **from the left hand** (clear F20):

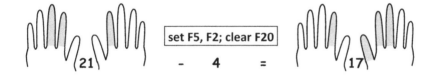

set F5, F2; clear F20

- 4 =

leaving us with the answer **17**.

We will use the following facts:

$$- 4 = - 10 + 6 = + 6 - 10$$
$$- 3 = - 10 + 7 = + 7 - 10$$
$$- 2 = - 10 + 8 = + 8 - 10$$

Our preference is to add to the right hand first and then minus from the left hand. But some people prefer to do it the other way around so we won't make a rule about it. We note, however, that with a little practice, you should find yourself performing both operations *at the same time* so the question of which to do first does not arise.

10.1 Fast way to minus 4 when right hand has 0, 1, 2 or 3

When the right hand has a value of 3 or less, 4 cannot be subtracted from this hand directly. ***To subtract 4***, we must ***subtract 1 from the left hand*** (which, remember, has a value of 10) ***and add 6 to the right hand***. We have **- 10 + 6 = - 4** which is the same as **- 4 = + 6 - 10**.

Some people prefer to minus 10 from the left hand first and then add 6 to the right. Others prefer to add first and then minus. It does not matter too much since, in the end, you should reach to the stage where you can ***minus 10 and add 6 at the same time***.

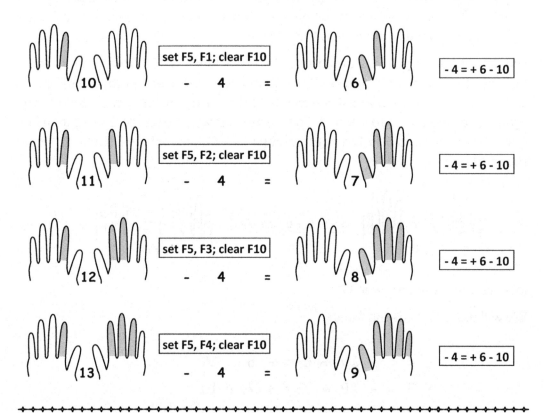

10	set F5, F1; clear F10 - 4 =	6 - 4 = + 6 - 10
11	set F5, F2; clear F10 - 4 =	7 - 4 = + 6 - 10
12	set F5, F3; clear F10 - 4 =	8 - 4 = + 6 - 10
13	set F5, F4; clear F10 - 4 =	9 - 4 = + 6 - 10

Practice: 10 – 4, 11 – 4, 12 – 4, 13 – 4, 14 – 4
15 – 4, 16 – 4, 17 – 4, 18 – 4, 19 – 4

> **Do Exercise 10.1, page 138**

10.2 Fast way to minus 3 when right hand has 0, 1 or 2

When the right hand has a value of 2 or less, 3 cannot be subtracted from this hand directly. **To subtract 3**, we must **subtract 1 from the left hand** (which, remember, has a value of 10) **and add 7 to the right hand**. We have **- 10 + 7 = - 3** which is the same as **- 3 = + 7 - 10**.

Some people prefer to minus 10 from the left hand first and then add 7 to the right. Others prefer to add first and then minus. It does not matter too much since, in the end, you should reach to the stage where you can **minus 10 and add 7 at the same time**.

(10)	set F5, F1-F2; clear F10 - 3 =	(7) - 3 = + 7 - 10
(11)	set F5, F2-F3; clear F10 - 3 =	(8) - 3 = + 7 - 10
(12)	set F5, F3-F4; clear F10 - 3 =	(9) - 3 = + 7 - 10

Practice: 10 – 3, 11 – 3, 12 – 3, 13 – 3, 14 – 3

           15 – 3, 16 – 3, 17 – 3, 18 – 3, 19 – 3

Do Exercise 10.2, page 138

10.3 Fast way to minus 2 when right hand has 0 or 1

When the right hand has a value of 0 or 1, 2 cannot be subtracted from this hand directly. **To subtract 2**, we must **subtract 1 from the left hand** (which, remember, has a value of 10) **and add 8 to the right hand**. We have **- 10 + 8 = - 2** which is the same as **- 2 = + 8 - 10**.

Some people prefer to minus 10 from the left hand first and then add 8 to the right. Others prefer to add first and then minus. It does not matter too much since, in the end, you should reach to the stage where you can **minus 10 and add 8 at the same time**.

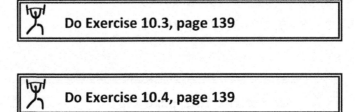

Practice: 10 – 2, 11 – 2, 12 – 2, 13 – 2, 14 – 2
15 – 2, 16 – 2, 17 – 2, 18 – 2, 19 – 2

Do Exercise 10.3, page 139

Do Exercise 10.4, page 139

11
Addition/subtraction: 2-digit numbers

In chapter 3, we showed you how to perform simple addition and subtraction of 2-digit numbers. The addition did not require any *carry* from the right hand to the left and the subtraction did not require us to *borrow* from the left hand.

Now that you know how to add and subtract single-digit numbers in all the different situations, we are ready to look at general problems which may require *carry* and/or *borrow*. We begin with addition.

11.1 Add 2-digit numbers with carry

We consider problems where we add a 2-digit number to a 2-digit number and there *is* carry from the right hand to the left hand. As usual, we will add *from left to right*. For example, consider

$$
\begin{array}{r}
17 \\
+18 \\
\hline
35
\end{array}
$$

Store **17** by storing **1** on the left hand and **7** on the right

Add **1** to the left hand, giving

set F20

Add **8** to the right hand, giving

clear F2, F1; set F30

Note that to add **8** to the right hand, you must minus **2** from the right hand (clear F2, F1) and add **1** (value 10) to the left hand (set F30).

$$
\begin{array}{r}
26 \\
+45 \\
\hline
71
\end{array}
$$

Store **26** by storing **2** on the left hand and **6** on the right

Add **4** to the left hand, giving

set F50; clear F20

Add **5** to the right hand, giving

clear F5; set F20

Note that to add **5** to the right hand, you must minus **5** from the right hand (clear F5) and add **1** (value 10) to the left hand (set F20).

$$\begin{array}{r} 36 \\ +47 \\ \hline 83 \end{array}$$

Store **36** by storing **3** on the left hand and **6** on the right

Add **4** to the left hand, giving

set F50; clear F30

Add **7** to the right hand, giving

clear F5; set F2-F3,F30

Note that to add **7** to the right hand, you minus **3** from the right hand (clear F5, set F2-F3) and add **1** (value 10) to the left hand (set F30).

+-+

$$\begin{array}{r} 28 \\ +56 \\ \hline 84 \end{array}$$

Store **28** by storing **2** on the left hand and **8** on the right

Add **5** to the left hand, giving

set F50

Add **6** to the right hand, giving

clear F5; set F4, F30

Note that to add **6** to the right hand, you must minus **4** from the right hand (clear F5, set F4) and add **1** (value 10) to the left hand (set F30).

+-+

You can use this method to add more than two 2-digit numbers, provided the answer is not more than 99. You simply add the next number to the current value stored on your hands.

Consider

$$17 +$$
$$18$$
$$23$$
$$25$$
$$\overline{83}$$

We add each number, in turn.

Store **17** by storing **1** on the left hand and **7** on the right

Add **18** to this:

Add **1** to the left hand, giving

> set F20

Add **8** to the right hand, giving

> clear F2, F1; set F30

Add **23** to this:

Add **2** to the left hand, giving

> set F50; clear F30-F10

Add **3** to the right hand, giving

> set F1-F3

Add **25** to this:

Add **2** to the left hand, giving

> set F10-F20

Add **5** to the right hand, giving

> clear F5; set F30

 Do Exercise 11.1, page 140

11.2 Subtract 2-digit numbers with borrow

We consider problems where we subtract from a 2-digit number and we need to borrow from the 10s column (left hand). For example, consider

$$
\begin{array}{r}
32 \\
-15 \\
\hline
17
\end{array}
$$

Store **32** by storing **3** on the left hand and **2** on the right

Subtract **1** from the left hand, giving

clear F30

Subtract **5** from the right hand, giving

set F5; clear F20

Note that to subtract **5** from the right hand, you must add **5** to the right hand (set F5) and minus **1** (value 10) from the left hand (clear F20).

$$
\begin{array}{r}
85 \\
-29 \\
\hline
56
\end{array}
$$

Store **85** by storing **8** on the left hand and **5** on the right

Subtract **2** from the left hand, giving

clear F30-F20

Subtract **9** from the right hand, giving

set F1; clear F10

Note that to subtract **9** from the right hand, you must add **1** to the right hand (set F1) and minus **1** (value 10) from the left hand (clear F10).

$$\begin{array}{r} 63 \\ -27 \\ \hline 36 \end{array}$$

Store **63** by storing **6** on the left hand and **3** on the right

Subtract **2** from the left hand, giving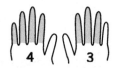

clear F50; set F20-F40

Subtract **7** from the right hand, giving

set F5; clear F3,F2,F40

Note that to subtract **7** from the right hand, you must add **3** to the right hand (set F5; clear F3-F2) and minus **1** (value 10) from the left hand (clear F40).

$$\begin{array}{r} 91 \\ -36 \\ \hline 55 \end{array}$$

Store **91** by storing **9** on the left hand and **1** on the right

Subtract **3** from the left hand, giving

clear F40-F20

Subtract **6** from the right hand, giving

set F5; clear F1, F10

Note that to subtract **6** from the right hand, you must add **4** to the right hand (set F5; clear F1) and minus **1** (value 10) from the left hand (clear F10).

Do Exercise 11.2, page 141

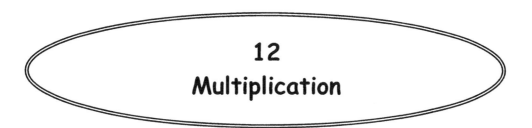

12
Multiplication

n this chapter, we show you how to multiply using your hands.

First we'll show you how you can learn your multiplication tables by doing repeated addition.

Then, we explain how you can use your tables to perform 2-digit by 1-digit multiplication.

As an added bonus, we show you an ingenious technique to do the 'bigger' table entries (like **9 x 8**) quickly and accurately.

12.1 Multiplication 1 x 1

You can use your hands to learn your multiplication tables. All you need to remember is that multiplication is just shorthand for repeated addition. For instance

4 x 2 is shorthand for 2 + 2 + 2 + 2 (we normally say "four 2s")

Also, since **4 x 2 = 2 x 4** (two 4s), we can write

4 x 2 = 2 x 4 = 2 + 2 + 2 + 2 = 4 + 4

So, to perform **4 x 2** (four 2s), we can add 2 four times, like this:

Set 2 and count "1"

Add 2 and count "2"

Add 2 and count "3"

Finally,

Add 2 and count "4"

So **4 x 2** is **8**.

Similarly, we can set 4 and count "1"

Next, add 4 and count "2"

So **2 x 4** is **8**.

To do **4 x 5** (four 5s)

Set 5 and count "1"

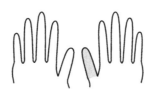

Add 5 and count "2"

Add 5 and count "3"

Add 5 and count "4"

So **4 x 5** is **20**.

Let's try **6 x 8** (six 8s)

Set 8 and count "1"

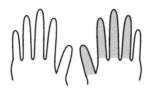

Add 8 and count "2"

Add 8 and count "3"

Add 8 and count "4"

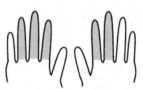

Add 8 and count "5"

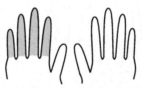

Add 8 and count "6"

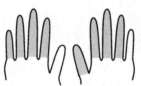

So **6 x 8** is **48**.

How about **7 x 9** (seven 9s)?

Set 9 and count "1"

Add 9 and count "2"

Add 9 and count "3"

Add 9 and count "4"

Add 9 and count "5"

Add 9 and count "6"

Add 9 and count "7"

So **7 x 9** is **63**.

Exercises

Do the following, counting 1 as you set the first, and counting 2, 3, 4, 5, 6, 7, 8, 9 as you add each of the others.

2 + 2 + 2 + 2 + 2 + 2 + 2 + 2 + 2

3 + 3 + 3 + 3 + 3 + 3 + 3 + 3 + 3

4 + 4 + 4 + 4 + 4 + 4 + 4 + 4 + 4

5 + 5 + 5 + 5 + 5 + 5 + 5 + 5 + 5

6 + 6 + 6 + 6 + 6 + 6 + 6 + 6 + 6

7 + 7 + 7 + 7 + 7 + 7 + 7 + 7 + 7

8 + 8 + 8 + 8 + 8 + 8 + 8 + 8 + 8

9 + 9 + 9 + 9 + 9 + 9 + 9 + 9 + 9

If you know your tables, that's fine. But if you have trouble learning them, this is a sure way to get the right answer. It may take you a little longer than if you knew the answer by heart but, at least, you'll get the right answer soon enough. Also, if you practice regularly, you'll be able to find the answers fairly quickly and, eventually, you will know your tables by heart.

 Do Exercise 12.1, page 142

12.2 Multiplication 2 × 1

Once you know your tables, *you can perform any 2-digit by 1-digit multiplication using your hands*. The product of a 2 × 1 multiplication is at most 3 digits long. The biggest such problem is 99 × 9 which is 891, so no 2 × 1 multiplication can have an answer bigger than 891. We note also that some 2 × 1 multiplications would be only 2 digits long, for example, 24 x 3 = 72.

We will use the left hand to hold the "*hundreds*" digit of the answer and the right hand to hold the "*tens*" digit. We will not store the "*units*" digit but simply call it out after the "hundreds" and "tens" digits.

Consider

$$34 \times 7 = 238$$

As mentioned before, in DigitalMath (and mental arithmetic, in general), it is faster working from left to right. Begin with **3 x 7 = 21**. Store **2** on the *left* hand and **1** on the *right*:

Next, multiply **4 x 7 = 28**. Add **2 *to the right hand*** and ***remember* 8**:

Read off the answer from left to right: **2 3 8**.

$$43 \times 8 = 344$$

Begin with **4 x 8 = 32**. Store **3** on the left hand and **2** on the right:

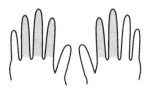

Next, multiply **3 x 8 = 24**. Add **2 *to the right hand*** and ***remember* 4**:

Read off the answer from left to right: **3 4 4**.

$$83 \times 6 = 498$$

Begin with **8 x 6 = 48**. Store **4** on the left hand and **8** on the right:

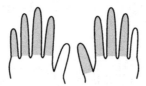

Next, multiply **3 x 6 = 18**. Add **1** *to the right hand* and *remember* **8**;

Read off the answer from left to right: **4 9 8**.

<hr>

$$78 \times 8 = 624$$

Begin with **7 x 8 = 56**. Store **5** on the left hand and **6** on the right:

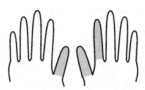

Next, multiply **8 x 8 = 64**. Add **6** *to the right hand* and *remember* **4**;

Read off the answer from left to right: **6 2 4**.

In all the examples so far, each 1-digit by 1-digit resulted in a 2-digit answer, so the process was straightforward. But what about **62 x 4**?

Here, when you multiply **2 x 4**, you get **8**. So what do you add to the right hand? **8** or **0**? Zero, of course, but you must be careful.

In order to avoid these kinds of problems, we will make the following rule:

*Multiplying a 1-digit number by a 1-digit number **always** gives a 2-digit answer*

If the answer *is* a 1-digit number (as in **2 x 4 = 8**), we will put **0** in front of it and say **08**. So **2 x 3 = 06, 1 x 7 = 07** and **3 x 3 = 09**.

If you remember to follow this rule, you should not have any problems with 2 x 1 multiplication. Let us show how to use the rule with some examples.

| 62 x 4 = 248 |

Begin with **6 x 4 = 24**. Store **2** on the left hand and **4** on the right:

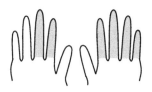

Next, multiply **2 x 4 = 08**. Add **0** to the right hand (do nothing) and **remember 8**. You have

so the answer is **2 4 8**.

$$71 \times 9 = 639$$

Begin with **7 x 9 = 63**. Store **6** on the left hand and **3** on the right:

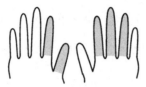

Next, multiply **1 x 9 = 09**. Add **0** to the right hand (do nothing) and *remember* **9**. You have

so the answer is **6 3 9**.

$$47 \times 2 = 94$$

Start with **4 x 2 = 08**. Store **0** on the left hand and **8** on the right:

Next, multiply **7 x 2 = 14**. Add *1 to the right hand* and *remember* **4**:

The answer is **0 9 4** or, simply, **94**.

19 x 8 = 152

Start with **1 x 8 = 08**. Store **0** on the left hand and **8** on the right:

Next, multiply **9 x 8 = 72**. Add **7 *to the right hand*** and ***remember* 2**:

The answer is **1 5 2**.

One final example:

85 x 8 = 680

Begin with **8 x 8 = 64**. Store **6** on the left hand and **4** on the right:

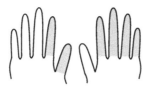

Next, multiply **5 x 8 = 40**. Add **4 *to the right hand*** and ***remember* 0**:

Read off the answer from left to right: **6 8 0**.

 Do Exercise 12.2, page 142

12.3 A faster way to do tables

Many children, and some adults, have difficulty with bigger values in the multiplication tables, such as 9 x 8, 8 x 7 and 7 x 9. In this section, we show you a faster way to do the 'bigger' tables. The method works from 6 x 6 up to 9 x 9. (If needed, use the method described in section 12.1 for 'smaller' tables.)

For this method, we think of the left hand as storing the values 1 to 9 rather than 10 to 90. We illustrate the method with **8 x 7**.

1. *Store 8* on the left hand and **7** on the right hand:

2. *Minus 1* from each hand: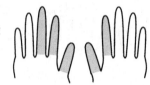

3. *Count the number of fingers* that are set. There are 3 on the left hand and 2 on the right, giving a total of **5**. *Multiply* this by 10, giving **50**.

4. *Multiply* the number of 'clear' fingers on the left hand by the number of 'clear' fingers on the right hand. This is **2 x 3**, giving **6**.

5. *Add* the values from 3 and 4: **50 + 6 = 56**.

6. Thus **8 x 7 = 56**.

+·+

Let's try **9 x 6**

1. *Store 9* on the left hand and **6** on the right hand: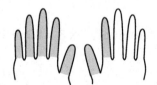

2. *Minus 1* from each hand: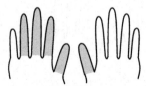

3. *Count the number of fingers* that are set. There are 4 on the left hand and 1 on the right, giving a total of **5**. *Multiply* this by 10, giving **50**.

4. *Multiply* the number of 'clear' fingers on the left hand by the number of 'clear' fingers on the right hand. This is **1 x 4**, giving **4**.

5. *Add* the values from 3 and 4: **50 + 4 = 54**, the answer to **9 x 6**.

How about **7 x 6?**

1. *Store* **7** on the left hand and **6** on the right hand:

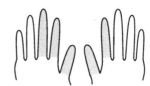

2. *Minus 1* from each hand:

3. A total of **3** fingers are now set. *Multiply* this by 10, giving **30**.

4. *Multiply* the number of 'clear' fingers on the left hand (**3**) by the number of 'clear' fingers on the right hand (**4**). This is **3 x 4 = 12**.

5. *Add* the values from 3 and 4: **30 + 12 = 42**. We have **7 x 6 = 42**.

✦✦

In practice, the calculation goes much faster than the above explanation might suggest. Once you've set the numbers and minus 1, you can read off the answer.

We note that, except for **6 x 6** and **7 x 6**, the number of set fingers gives the first digit of the answer, and multiplying the 'clear' fingers of one hand by the 'clear' fingers of the other hand gives the second digit.

For example, for **7 x 9**, after we set 7 on the left, 9 on the right, and minus 1 from both hands, we have:

There are **6** set fingers so the first digit is **6**. Multiplying **3** (number of clear fingers on the left) by **1** (number of clear fingers on the right) gives us **3**, so the answer is **63**.

✦✦

Exercises: Use the method of this section to do the following:

9 x 8	7 x 7	8 x 8	9 x 7	6 x 6
7 x 8	9 x 9	8 x 9	6 x 8	8 x 6
7 x 9	6 x 7	6 x 9	8 x 7	9 x 6

Exercises

Exercise 1.1: Set and recognise numbers 0 to 9 (right hand)

Set your fingers to each of the following, say the number, then write the number on the hand

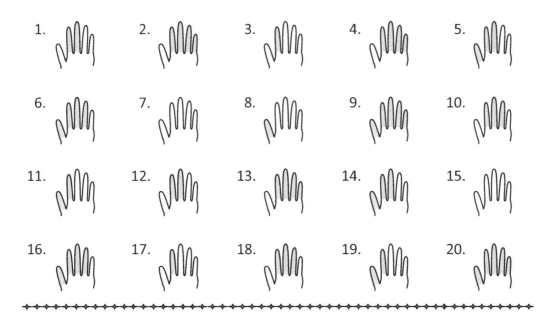

Exercise 1.2: Count from 0 to 9

Drill: count from 0 to 9 and back to 0 on right hand

Goal: to be able to recognize the value represented by any finger position and get used to adding and taking away fingers of the right hand

Count the following:

1. **0-1-0**
2. **0-1-2-1-0**
3. **0-1-2-3-2-1-0**
4. **0-1-2-3-4-3-2-1-0**
5. **0-1-2-3-4-5-4-3-2-1-0**
6. **0-1-2-3-4-5-6-5-4-3-2-1-0**
7. **0-1-2-3-4-5-6-7-6-5-4-3-2-1-0**
8. **0-1-2-3-4-5-6-7-8-7-6-5-4-3-2-1-0**
9. **0-1-2-3-4-5-6-7-8-9-8-7-6-5-4-3-2-1-0**

Exercise 1.3: Set and recognise numbers 0 to 90 (left hand)

Set your fingers to each of the following, say the number, then write the number on the hand

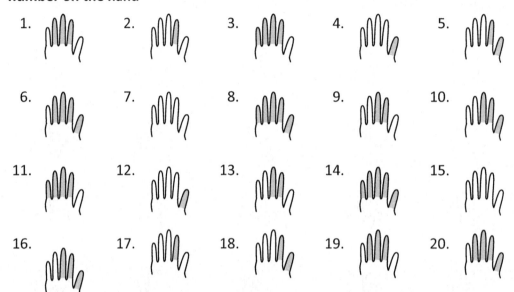

+++

Exercise 1.4: Count from 10 to 90

Drill: count in 10s from 0 to 90 and back to 0 on left hand

Goal: to be able to recognize the value represented by any finger position and get used to adding and taking away fingers of the left hand

Count the following:

1. `0-10-0`
2. `0-10-20-10-0`
3. `0-10-20-30-20-10-0`
4. `0-10-20-30-40-30-20-10-0`
5. `0-10-20-30-40-50-40-30-20-10-0`
6. `0-10-20-30-40-50-60-50-40-30-20-10-0`
7. `0-10-20-30-40-50-60-70-60-50-40-30-20-10-0`
8. `0-10-20-30-40-50-60-70-80-70-60-50-40-30-20-10-0`
9. `0-10-20-30-40-50-60-70-80-90-80-70-60-50-40-30-20-10-0`

Exercise 1.5: Count from 0 to 99

Drill: count from 0 to 99

Drill: count from 99 down to 0

Goal: to be able to recognize the value represented by any finger position and get used to adding and taking away fingers from both hands

Set your fingers to each of the following, say the number, then write the number on the hands.

Exercise 2.1: Addition drill using F1-F4

Do the following by counting. Repeat without counting.

1. 1 + 1	2. 1 + 3	3. 2 + 1	4. 2 + 2
5. 3 + 1	6. 1 + 2 + 1	7. 2 + 1 + 1	8. 1 + 1 + 2

Exercise 2.2: Subtraction drill using F1-F4

Do the following by counting. Repeat without counting.

1. 2 − 1	2. 2 − 2	3. 3 − 1	4. 3 − 2
5. 3 − 3	6. 4 − 1	7. 4 − 3	8. 4 − 4

Exercise 2.3: Addition and subtraction using F1-F4

Do the following by counting. Repeat without counting.

1	2	3	4	5	6	7	8	9	10
1	1	1	1	1	1	1	1	1	2
1	1	2	2	2	3	3	3	3	1
−1	−2	−1	−2	−3	−1	−2	−3	−4	−1

11	12	13	14	15	16	17	18	19	20
2	2	2	2	2	2	3	3	3	3
1	1	2	2	2	2	1	1	1	1
−2	−3	−1	−2	−3	−4	−1	−2	−3	−4

21	22	23	24	25	26	27	28	29	30
2	1	3	1	4	4	4	3	2	3
2	3	1	2	−3	−2	−1	−1	1	−2
−3	−2	−1	−3	2	1	−2	2	1	3
2	1	−1	4	−1	−3	3	−3	−3	−2
1	1	2	−2	2	2	−4	2	2	1

Exercise 2.4: Addition drill with F1 to F5

Do the following by counting. Repeat without counting.

1. 1 + 5   2. 2 + 5   3. 3 + 5   4. 4 + 5

5. 5 + 1   6. 5 + 2   7. 5 + 3   8. 5 + 4

+‑+

Exercise 2.5: Basic addition with F1 to F5

Do the following by counting. Repeat without counting.

1	2	3	4	5	6	7	8	9	10
1	5	2	3	2	1	1	1	2	5
5	3	5	5	2	5	3	5	5	2
2	1	2	1	5	1	5	3	1	2

11	12	13	14	15	16	17	18	19	20
5	1	5	5	3	1	2	5	6	2
1	2	1	2	1	1	1	1	1	6
1	5	3	1	5	5	5	2	1	1

21	22	23	24	25	26	27	28	29	30
1	2	6	1	7	1	6	1	1	1
6	1	2	2	1	7	1	1	6	1
2	6	1	6	1	1	2	7	1	6

31	32	33	34	35	36	37	38	39	40
2	1	6	2	1	1	1	1	1	5
1	1	1	1	5	2	1	1	5	2
5	5	1	1	1	5	6	5	2	1
1	2	1	5	1	1	1	1	1	1

Repeat these exercises without looking at your hand.

Exercise 2.6: Subtraction drill with F1 to F5

Do the following by counting. Repeat without counting.

1. 9 − 3	2. 9 − 5	3. 8 − 3	4. 7 − 5	5. 8 − 5
6. 9 − 6	7. 8 − 7	8. 6 − 5	9. 9 − 7	10. 7 − 6

✦✦

Exercise 2.7: Basic subtraction with F1 to F5

1	2	3	4	5	6	7	8	9	10
3	6	8	9	3	7	8	9	2	8
5	2	−5	−4	6	−2	−6	−1	5	−7
−2	−5	1	2	−5	−5	5	−5	−6	8

11	12	13	14	15	16	17	18	19	20
2	6	9	9	2	8	3	6	2	9
5	3	−6	−8	6	−1	5	−5	5	−7
−3	−4	5	6	−7	−5	−6	7	−6	6

21	22	23	24	25	26	27	28	29	30
5	2	4	5	4	5	9	3	2	6
4	5	−3	1	5	4	−4	5	2	2
−3	−1	5	3	−3	−6	3	−6	5	−5
2	2	−1	−5	1	−1	−5	2	−3	−1

31	32	33	34	35	36	37	38	39	40
8	7	4	3	1	5	3	5	1	1
−2	2	5	5	5	4	−1	1	3	2
3	−5	−6	−6	3	−3	5	3	−2	5
−5	−3	5	7	−7	2	−2	−6	5	−8

(Exercise 2.7 continued: subtraction with F1-F5)

41	42	43	44	45	46	47	48	49	50
3	4	3	1	7	6	9	2	4	8
5	5	5	3	−2	−5	−6	5	−1	−5
−2	−6	1	−2	4	2	5	−6	6	1
3	5	−4	5	−3	6	−7	7	−8	−3
−6	−3	3	−6	1	−7	3	−2	7	6

Repeat these exercises without looking at your hand.

+-+

Exercise 2.8: Addition drill with a switch from 4 to 5

Do the following by counting.

1. 2 + 3   2. 4 + 1   3. 3 + 3   4. 4 + 2   5. 2 + 4

6. 3 + 4   7. 4 + 3   8. 1 + 4   9. 4 + 4   10. 3 + 2

+-+

Exercise 2.9: Problems which require a switch from 4 to 5

1	2	3	4	5	6	7	8	9	10
1	3	2	3	1	4	2	1	2	4
2	1	4	3	4	1	2	1	4	3
3	2	2	3	3	2	2	4	1	2

11	12	13	14	15	16	17	18	19	20
4	3	2	2	4	1	3	3	4	2
−2	3	4	3	4	3	1	4	−1	3
3	−5	−1	2	−3	3	1	−5	4	4

(Exercise 2.9 continued: switch from 4 to 5)

21	22	23	24	25	26	27	28	29	30
6	7	8	9	9	9	8	7	4	7
−5	−2	−6	−5	−6	−7	−5	−6	4	−5
4	3	3	4	3	4	3	4	−5	4

31	32	33	34	35	36	37	38	39	40
1	7	8	9	9	9	4	4	4	9
2	−5	−6	−6	−4	−8	4	3	4	−7
6	3	4	2	3	4	−6	−7	−3	3

41	42	43	44	45	46	47	48	49	50
8	7	4	3	5	5	3	6	8	1
−5	2	5	5	3	4	6	−5	−6	2
3	−5	−6	−6	−6	−7	−7	2	2	3
2	3	4	2	3	4	3	4	2	2

❖❖

Exercise 2.10: Subtraction drill with a switch from 5 to 4

Do the following by counting.

1. $5 - 1$ 2. $6 - 2$ 3. $7 - 3$ 4. $8 - 4$ 5. $9 - 5$

6. $7 - 4$ 7. $6 - 3$ 8. $8 - 6$ 9. $9 - 6$ 10. $6 - 4$

❖❖

Exercise 2.11: Problems which require a switch from 5 to 4

1	2	3	4	5	6	7	8	9	10
5	5	7	8	6	6	2	1	4	9
2	−2	−4	−6	−3	−4	4	4	3	−7
−3	4	3	4	2	6	−3	−2	−4	4

11	12	13	14	15	16	17	18	19	20
6	6	7	7	3	8	8	9	9	5
−2	−3	−5	−3	4	−5	−4	−6	−8	−3
3	5	3	4	−3	2	1	4	4	6

21	22	23	24	25	26	27	28	29	30
4	1	2	3	1	3	2	6	2	5
4	6	7	5	5	3	5	2	7	−4
−4	−3	−5	−4	−2	−3	−4	−4	−5	7

31	32	33	34	35	36	37	38	39	40
4	7	8	9	6	6	9	8	8	8
3	−4	−4	−6	−4	−5	−6	−5	−6	−5
−4	2	3	2	3	4	3	4	3	3

41	42	43	44	45	46	47	48	49	50
8	7	4	7	9	8	6	6	9	4
−5	2	5	−4	−6	−7	−3	−4	−7	4
3	−3	−4	3	5	4	4	5	6	−2
−2	−3	−2	−2	−4	−1	−3	−4	−4	−2

Exercise 3.1: Add 1-digit to 2-digit without carry

1	2	3	4	5	6	7	8	9	10
11	21	17	15	23	30	26	25	42	32
3	5	2	4	4	7	3	4	6	7

11	12	13	14	15	16	17	18	19	20
75	62	44	15	83	31	35	65	71	34
4	6	4	4	3	7	3	4	8	5

21	22	23	24	25	26	27	28	29	30
33	41	92	53	74	35	73	82	54	84
4	7	7	2	3	2	5	6	5	4

31	32	33	34	35	36	37	38	39	40
42	63	52	55	23	91	94	64	46	33
5	4	3	3	6	7	5	3	3	4

Exercise 3.2: Add 2-digit to 2-digit without carry

1	2	3	4	5	6	7	8	9	10
11	21	17	15	23	30	27	25	42	36
22	13	12	24	25	15	51	23	50	52

11	12	13	14	15	16	17	18	19	20
25	32	44	15	83	31	37	65	37	34
32	36	53	64	12	46	30	23	42	61

(Exercise 3.2 continued)

21	22	23	24	25	26	27	28	29	30
33	41	28	23	47	14	73	22	45	38
44	24	40	32	42	42	13	36	54	41

31	32	33	34	35	36	37	38	39	40
24	16	30	15	26	24	53	42	35	38
52	33	56	52	61	24	14	54	23	50

Exercise 3.3: Subtract 1-digit from 2-digit without borrow

1	2	3	4	5	6	7	8	9	10
23	46	17	25	48	45	78	49	98	88
−1	−5	−2	−4	−5	−1	−6	−7	−3	−8

11	12	13	14	15	16	17	18	19	20
77	68	97	79	95	77	67	88	79	95
−4	−7	−7	−4	−3	−6	−5	−7	−6	−1

21	22	23	24	25	26	27	28	29	30
75	65	68	55	89	56	86	58	99	79
−2	−3	−5	−5	−8	−3	−4	−4	−4	−5

31	32	33	34	35	36	37	38	39	40
76	49	86	67	87	48	67	96	58	88
−2	−3	−4	−6	−5	−3	−3	−4	−6	−2

Exercise 3.4: Subtract 2-digit numbers without borrow

1	2	3	4	5	6	7	8	9	10
22	23	17	25	48	45	78	49	92	88
−11	−13	−12	−14	−25	−15	−51	−23	−50	−52

11	12	13	14	15	16	17	18	19	20
77	68	97	79	95	77	67	88	79	95
−32	−36	−53	−64	−12	−46	−30	−23	−42	−61

21	22	23	24	25	26	27	28	29	30
75	65	68	55	89	56	86	58	99	79
−44	−24	−40	−32	−42	−42	−13	−36	−54	−41

31	32	33	34	35	36	37	38	39	40
76	49	86	67	87	48	67	96	58	88
−52	−33	−56	−52	−61	−24	−14	−54	−23	−50

41	42	43	44	45	46	47	48	49	50
71	50	93	42	43	64	65	83	44	45
6	8	4	6	43	4	4	3	13	40
−25	−34	−61	−16	−36	−27	−38	−65	−41	−52

Exercise 3.5: Add/subtract 1- and 2-digit numbers without borrow

1	2	3	4	5	6	7	8	9	10
22	41	17	27	48	52	73	49	95	88
6	7	31	−4	−6	36	5	−23	−44	−52
−15	−23	−12	33	27	−18	−24	64	26	13

11	12	13	14	15	16	17	18	19	20
76	68	97	79	96	77	62	83	79	85
−22	−35	−26	−7	−12	−36	6	−33	−38	−41
34	52	7	13	4	16	−27	37	43	4

21	22	23	24	25	26	27	28	29	30
75	68	52	53	89	56	87	55	99	79
−4	−26	16	3	−64	−32	−13	32	−67	−40
23	7	−4	−21	22	3	23	−4	14	−6
−62	35	−13	42	−17	−6	−55	16	−2	12

31	32	33	34	35	36	37	38	39	40
55	47	25	42	98	76	23	54	79	84
21	−26	33	6	−53	−32	5	−12	−60	−13
13	55	−15	−35	4	54	61	36	−8	26
−57	−21	36	61	−27	−66	−26	−18	37	−90

Exercise 4.1: Basic drills in general single-digit addition

Do the following by counting. Repeat, doing each as fast as you can.

9 + 1 = 10	6 + 5 = 11	5 + 6 = 11
7 + 9 = 16	3 + 9 = 12	5 + 8 = 13
4 + 7 = 11	4 + 6 = 10	6 + 4 = 10
7 + 8 = 15	5 + 9 = 14	4 + 8 = 12
9 + 7 = 16	9 + 2 = 11	2 + 9 = 11
8 + 9 = 17	9 + 8 = 17	9 + 9 = 18
1 + 9 = 10	2 + 8 = 10	3 + 7 = 10
3 + 8 = 11	4 + 9 = 13	5 + 7 = 12
6 + 6 = 12	6 + 7 = 13	6 + 8 = 14
6 + 9 = 15	7 + 6 = 13	7 + 5 = 12
8 + 8 = 16	8 + 2 = 10	8 + 3 = 11
8 + 7 = 15	8 + 5 = 13	8 + 6 = 14
7 + 4 = 11	9 + 6 = 15	9 + 5 = 14
9 + 3 = 12	9 + 4 = 13	7 + 3 = 10

Exercise 4.2: Continuous drills in single-digit addition

Drills in addition: say each number after each addition

- Starting with 1, **add 2 repeatedly** until you reach 99.
- Starting with 2, **add 2 repeatedly** until you reach 98
- Starting with 5, **add 5 repeatedly** until you reach 95.
- Starting with 6, 7, 8 and 9, **add 5 repeatedly** as much as you can.
- Starting with 0, 1 and 2, **add 3 repeatedly** as much as you can.
- Starting with 0, 1, 2 and 3, **add 4 repeatedly** as much as you can.
- Starting with 0, 1, 2, 3, 4 and 5, **add 6 repeatedly** as much as you can.
- Starting with 0, 1, 2, 3, 4, 5 and 6, **add 7 repeatedly** as much as you can.
- Starting with 0, 1, 2, 3, 4, 5, 6 and 7, **add 8 repeatedly** as much as you can.
- Starting with 0, 1, 2, 3, 4, 5, 6, 7 and 8, **add 9 repeatedly** as much as you can.

Exercise 4.3: Drill problems in single-digit addition

The first time, do these problems by counting. Then do them again after you have learnt the fast way to add.

1	2	3	4	5	6	7	8	9	10
6	3	5	5	1	7	2	4	4	1
2	3	2	4	4	1	1	2	4	5
2	4	3	1	5	2	7	4	2	4

11	12	13	14	15	16	17	18	19	20
3	6	4	7	2	5	2	4	9	3
4	2	6	1	4	3	1	5	1	6
4	3	1	3	5	3	8	2	1	2

21	22	23	24	25	26	27	28	29	30
3	6	4	7	2	5	2	5	9	3
4	2	6	1	4	4	1	5	2	6
5	4	2	4	6	3	9	2	1	3

31	32	33	34	35	36	37	38	39	40
4	6	4	7	3	5	2	5	8	3
4	3	6	2	4	4	2	6	2	7
5	4	3	4	6	4	9	2	3	3

41	42	43	44	45	46	47	48	49	50
4	7	4	8	3	5	3	6	8	3
5	3	6	2	4	4	2	2	3	9
5	4	4	4	7	5	9	6	3	2

(Exercise 4.3 continued: single-digit addition)

51	52	53	54	55	56	57	58	59	60
5	8	4	9	3	6	3	6	7	4
5	3	5	2	4	4	3	2	5	9
5	4	6	4	8	5	9	7	3	2

61	62	63	64	65	66	67	68	69	70
5	8	4	9	4	6	4	7	8	4
5	4	5	3	4	4	3	2	5	9
6	4	7	4	8	6	9	7	3	3

71	72	73	74	75	76	77	78	79	80
6	8	4	9	6	7	4	7	9	5
5	4	5	4	4	3	4	6	5	5
6	5	8	4	7	7	9	4	3	7

81	82	83	84	85	86	87	88	89	90
6	8	4	9	6	8	6	7	9	5
6	5	5	4	5	6	4	4	6	6
6	5	9	5	7	4	8	7	3	7

91	92	93	94	95	96	97	98	99	100
6	8	5	9	7	8	6	3	9	5
6	5	5	3	5	7	5	8	6	6
7	6	9	7	7	4	8	8	4	8

Exercise 4.4: General problems in single-digit addition

The first time, do these problems by counting. Then do them again after you have learnt the fast way to add.

1	2	3	4	5	6	7	8	9	10
2	2	3	4	5	6	7	4	5	6
2	3	4	5	6	7	8	3	4	5
5	4	5	6	7	8	9	2	3	4

11	12	13	14	15	16	17	18	19	20
7	9	7	7	3	5	5	5	9	3
6	8	6	1	8	3	1	5	4	6
5	7	1	4	5	7	8	2	5	8

21	22	23	24	25	26	27	28	29	30
7	6	4	7	7	5	2	5	9	7
4	1	6	3	4	4	3	7	2	6
5	4	5	4	6	9	9	5	4	8

31	32	33	34	35	36	37	38	39	40
8	6	6	7	3	5	4	5	9	8
8	9	6	3	5	4	3	6	3	7
8	8	3	6	6	7	9	5	4	8

41	42	43	44	45	46	47	48	49	50
9	4	9	8	5	5	3	4	8	4
1	6	9	6	7	7	3	9	5	9
9	6	2	6	7	8	9	9	8	7

Exercise 4.5: Drills in general single-digit subtraction

Do the following by counting. Repeat, doing each as fast as you can.

10 − 1 = 9	11 − 5 = 6	11 − 6 = 5
16 − 9 = 7	12 − 3 = 9	13 − 5 = 8
11 − 4 = 7	10 − 4 = 6	10 − 6 = 4
15 − 7 = 8	14 − 5 = 9	12 − 4 = 8
16 − 7 = 9	11 − 2 = 9	11 − 9 = 2
17 − 8 = 9	17 − 9 = 8	18 − 9 = 9
10 − 9 = 1	10 − 2 = 8	10 − 3 = 7
11 − 3 = 8	13 − 4 = 9	12 − 5 = 7
12 − 6 = 6	13 − 6 = 7	14 − 6 = 8
15 − 6 = 9	13 − 7 = 6	12 − 7 = 5
16 − 8 = 8	10 − 8 = 2	11 − 8 = 3
15 − 8 = 7	13 − 8 = 5	14 − 8 = 6
11 − 7 = 4	15 − 9 = 6	14 − 9 = 5
12 − 9 = 3	13 − 9 = 4	10 − 7 = 3

Exercise 4.6: Continuous drills in single-digit subtraction

Drills in subtraction: say each number after each subtraction

- Start with 99; **minus 2 repeatedly** until you reach 1.
- Start with 98; **minus 2 repeatedly** until you reach 0.
- Start with 95; **minus 5 repeatedly** until you reach 0.
- Start with 99, 98, 97 and 96; **minus 5 repeatedly** as much as you can.
- Start with 99, 98 and 97; **minus 3 repeatedly** as much as you can.
- Start with 99, 98, 97 and 96; **minus 4 repeatedly** as much as you can.
- Start with 99, 98, 97, 96, 95 and 94; **minus 6 repeatedly** as much as you can.
- Start with 99, 98, 97, 96, 95, 94, 93; **minus 7 repeatedly** as much as you can.
- Start with 99, 98, 97, 96, 95, 94, 93, 92; **minus 8 repeatedly** as much as you can.
- Start with 99, 98, 97, 96, 95, 94, 93, 92 and 91; **minus 9 repeatedly** as much as you can.

Exercise 4.7: Drill exercises: subtraction from double–digit

1	2	3	4	5	6	7	8	9	10
10	10	10	10	10	10	10	10	10	10
−2	−3	−5	−4	−4	−1	−1	−2	−4	−5
−2	−4	−2	−1	−5	−2	−7	−4	−2	−4

11	12	13	14	15	16	17	18	19	20
11	11	11	11	11	11	11	11	11	11
−4	−2	−6	−1	−4	−3	−1	−5	−1	−6
−4	−3	−1	−3	−5	−3	−8	−2	−1	−2

21	22	23	24	25	26	27	28	29	30
12	12	12	12	12	12	12	12	12	12
−4	−2	−6	−1	−4	−4	−1	−5	−2	−6
−5	−4	−2	−4	−6	−3	−9	−2	−1	−3

31	32	33	34	35	36	37	38	39	40
13	13	13	13	13	13	13	13	13	13
−4	−3	−6	−2	−4	−4	−2	−6	−2	−7
−5	−4	−3	−4	−6	−4	−9	−2	−3	−3

41	42	43	44	45	46	47	48	49	50
14	14	14	14	14	14	14	14	14	14
−5	−3	−6	−2	−4	−4	−2	−2	−3	−9
−5	−4	−4	−4	−7	−5	−9	−6	−3	−2

(Exercise 4.7 continued: subtraction from double-digit)

51	52	53	54	55	56	57	58	59	60
15	15	15	15	15	15	15	15	15	15
−5	−3	−5	−2	−4	−4	−3	−2	−5	−9
−5	−4	−6	−4	−8	−5	−9	−7	−3	−2

61	62	63	64	65	66	67	68	69	70
16	16	16	16	16	16	16	16	16	16
−5	−4	−5	−3	−4	−4	−3	−2	−5	−9
−6	−4	−7	−4	−8	−6	−9	−7	−3	−3

71	72	73	74	75	76	77	78	79	80
17	17	17	17	17	17	17	17	17	17
−5	−4	−5	−4	−4	−3	−4	−6	−5	−5
−6	−5	−8	−4	−7	−7	−9	−4	−3	−7

81	82	83	84	85	86	87	88	89	90
18	18	18	18	18	18	18	18	18	18
−6	−5	−5	−4	−5	−6	−4	−4	−6	−6
−6	−5	−9	−5	−7	−4	−8	−7	−3	−7

91	92	93	94	95	96	97	98	99	100
19	19	19	19	19	19	19	19	19	19
−6	−5	−5	−3	−5	−7	−5	−8	−6	−6
−7	−6	−9	−7	−7	−4	−8	−8	−4	−8

Exercise 4.8: General subtraction from double-digit

1	2	3	4	5	6	7	8	9	10
9	9	12	15	18	21	24	9	12	12
−2	−3	−4	−5	−6	−7	−8	−3	−4	−5
−5	−4	−5	−6	−7	−8	−9	−2	−3	−4

11	12	13	14	15	16	17	18	19	20
18	24	14	12	16	15	14	12	18	17
−6	−8	−6	−1	−8	−3	−1	−5	−4	−6
−5	−7	−1	−4	−5	−7	−8	−2	−5	−8

21	22	23	24	25	26	27	28	29	30
16	11	15	14	17	18	14	17	15	21
−4	−1	−6	−3	−4	−4	−3	−7	−2	−6
−5	−4	−5	−4	−6	−9	−9	−5	−4	−8

31	32	33	34	35	36	37	38	39	40
24	23	15	16	14	16	16	16	16	23
−8	−9	−6	−3	−5	−4	−3	−6	−3	−7
−8	−8	−3	−6	−6	−7	−9	−5	−4	−8

41	42	43	44	45	46	47	48	49	50
19	16	20	20	19	20	15	22	21	20
−1	−6	−9	−6	−7	−7	−3	−9	−5	−9
−9	−6	−2	−6	−7	−8	−9	−9	−8	−7

Exercise 4.9: Miscellaneous problems: addition and subtraction

1	2	3	4	5	6	7	8	9	10
6	3	5	5	1	7	2	4	4	1
−2	3	2	−4	4	−1	−1	2	−4	5
2	−4	−3	1	−5	2	7	−4	2	−4

11	12	13	14	15	16	17	18	19	20
3	6	4	7	2	5	2	4	9	3
4	−2	6	1	4	−3	−1	5	1	6
−4	3	−1	−3	−5	3	8	−2	−1	−2

21	22	23	24	25	26	27	28	29	30
3	6	4	7	2	5	2	5	9	3
4	−2	6	1	4	−4	−1	5	−2	6
−5	4	−2	−4	−6	3	9	−2	1	−3

31	32	33	34	35	36	37	38	39	40
4	6	4	7	3	5	9	5	8	3
4	−3	6	−2	4	4	2	6	2	7
−5	4	−3	4	−6	−4	−9	−2	−3	−3

41	42	43	44	45	46	47	48	49	50
4	7	4	8	3	5	3	6	8	3
5	−3	6	−2	4	−4	9	−2	−3	−2
−5	4	−4	4	−7	5	−2	6	3	9

Exercise 4.10: Miscellaneous problems: addition and subtraction

1	2	3	4	5	6	7	8	9	10
5	8	4	9	3	6	9	6	7	4
5	3	5	−2	8	4	−3	−2	5	9
−5	−4	−6	4	−4	−5	4	7	−3	−2

11	12	13	14	15	16	17	18	19	20
5	8	4	9	4	6	4	7	8	4
5	−4	5	3	−4	−4	−3	2	−5	9
−6	4	−7	−4	8	6	9	−7	3	−3

21	22	23	24	25	26	27	28	29	30
6	8	4	9	6	7	4	7	9	5
5	−4	5	4	4	−3	−4	6	−5	5
−6	5	−8	−4	−7	7	9	−4	3	−7

31	32	33	34	35	36	37	38	39	40
6	8	4	9	6	8	6	7	9	5
6	−5	5	−4	−5	6	4	−4	6	6
−6	5	−9	5	7	−4	−8	7	−3	−7

41	42	43	44	45	46	47	48	49	50
6	8	5	9	7	8	6	3	9	5
−6	5	5	−3	5	7	−5	8	−6	6
7	−6	−9	7	−7	−4	8	−8	4	−8

Exercise 4.11: Miscellaneous problems: addition and subtraction

1	2	3	4	5	6	7	8	9	10
2	2	3	4	5	6	7	4	5	6
−2	3	4	5	6	7	8	−3	4	−5
5	−4	−5	−6	−7	−8	−9	2	−3	4

11	12	13	14	15	16	17	18	19	20
7	9	7	7	3	5	5	5	9	3
6	−8	6	−1	8	3	−1	5	−4	6
−5	7	−1	4	−5	−7	8	−2	5	−8

21	22	23	24	25	26	27	28	29	30
7	6	4	7	7	5	2	5	9	7
−4	1	6	3	−4	−4	9	7	−2	6
5	−4	−5	−4	6	9	−3	−5	4	−8

31	32	33	34	35	36	37	38	39	40
8	6	6	7	3	5	4	5	9	8
−8	9	6	−3	5	4	−3	6	−3	7
8	−8	−3	6	−6	−7	9	−5	4	−8

41	42	43	44	45	46	47	48	49	50
9	4	9	8	5	5	8	9	8	4
1	6	−9	6	7	7	3	−4	5	9
−9	−6	2	−6	−7	−8	−9	9	−8	−7

Exercise 5.1: Add 4 (+ 4 = + 5 − 1)

1	2	3	4	5	6	7	8	9	10
4	1	2	3	2	1	4	2	4	5
1	3	2	4	3	4	−2	4	−1	4
4	4	4	2	4	4	4	2	4	−3

11	12	13	14	15	16	17	18	19	20
9	8	5	8	4	7	6	3	9	6
−6	−5	−5	−6	4	−2	−5	4	−5	−3
4	4	4	4	−2	4	4	−5	4	4

21	22	23	24	25	26	27	28	29	30
8	8	4	3	5	9	7	7	8	9
−7	−5	4	4	4	−7	−5	−6	−8	−8
4	4	−5	−6	−6	4	4	4	4	4

Exercise 5.2: Add 3 (+ 3 = + 5 − 2)

1	2	3	4	5	6	7	8	9	10
3	1	2	5	2	1	4	2	4	6
3	3	2	1	4	4	−2	3	1	3
3	3	3	3	3	3	3	3	3	−4

11	12	13	14	15	16	17	18	19	20
9	8	5	8	4	7	6	4	9	6
−6	−5	−5	−6	3	−2	−5	3	−5	−3
3	3	3	3	−2	3	3	−5	3	3

(Exercise 5.2 continued)

21	22	23	24	25	26	27	28	29	30
8	8	3	3	5	9	7	7	8	9
−7	−5	3	3	3	−7	−5	−6	−8	−8
3	3	−5	−6	−6	3	3	3	3	3

Exercise 5.3: Add 2 (+ 2 = + 5 − 3)

1	2	3	4	5	6	7	8	9	10
2	1	3	5	2	1	4	4	3	6
2	3	2	1	4	4	−2	−1	3	2
2	2	3	2	2	2	2	2	2	−5

11	12	13	14	15	16	17	18	19	20
9	8	5	8	4	8	6	4	9	6
−6	−5	−5	−6	2	−3	−5	2	−5	−3
2	2	2	2	−1	2	2	−5	2	2

21	22	23	24	25	26	27	28	29	30
8	8	3	7	5	9	7	7	8	3
−7	−5	2	2	2	−7	−4	−6	−8	2
2	2	−5	−6	−6	2	2	2	2	−2

Exercise 5.4: Add 1 (+ 1 = + 5 − 4)

1	2	3	4	5	6	7	8	9	10
1	3	4	8	2	2	5	9	8	6
1	1	1	1	4	2	3	−5	−4	2
1	5	4	−3	1	1	1	1	1	1

11	12	13	14	15	16	17	18	19	20
5	3	8	4	2	8	6	4	7	1
4	6	1	1	7	−3	3	2	−3	8
1	1	1	3	1	1	1	−5	1	1

＋＋

Exercise 5.5: Add 5 (+ 5 = − 5 + 10)

1	2	3	4	5	6	7	8	9	10
2	1	3	4	5	7	2	4	1	5
5	3	5	5	3	5	3	2	5	4
2	5	3	4	5	3	5	5	3	5

11	12	13	14	15	16	17	18	19	20
4	8	5	8	9	7	7	6	9	8
−2	−1	5	5	−7	−2	5	5	−2	−6
5	5	−3	−2	5	5	−3	−3	5	5

21	22	23	24	25	26	27	28	29	30
2	5	3	6	9	7	1	4	8	8
5	5	5	5	5	5	5	5	5	5
5	5	5	5	5	5	5	5	5	6

Exercise 5.6: Revision: add 1, 2, 3, 4, 5

1	2	3	4	5	6	7	8	9	10
1	3	5	3	5	4	3	2	8	6
2	4	4	5	5	4	3	3	1	5
3	5	3	4	1	2	5	4	5	6

11	12	13	14	15	16	17	18	19	20
3	2	2	2	5	5	3	1	1	1
3	1	4	3	3	2	4	4	5	3
2	2	2	4	3	4	2	2	2	3
1	5	1	3	3	2	4	3	2	3

21	22	23	24	25	26	27	28	29	30
4	2	3	2	5	4	8	3	1	4
4	5	2	3	1	3	−3	4	5	4
2	−3	3	5	5	−5	5	5	2	−2
2	3	−1	−2	−2	1	4	−2	2	5

31	32	33	34	35	36	37	38	39	40
7	5	4	2	4	7	8	9	9	6
5	5	5	6	4	2	5	5	5	3
2	4	5	5	5	5	6	−3	5	−2
4	4	2	2	4	2	−4	4	5	5

41	42	43	44	45	46	47	48	49	50
5	4	3	2	5	4	9	7	8	6
5	4	3	3	3	3	5	2	1	5
5	5	3	4	1	5	4	5	5	3
5	3	5	5	1	5	1	2	3	1
5	5	1	5	4	2	1	2	5	4

Exercise 6.1: Minus 4 (– 4 = – 5 + 1)

1	2	3	4	5	6	7	8	9	10
6	1	7	9	2	4	7	9	8	9
3	3	2	–4	3	1	–2	–4	–1	–3
–4	–4	–4	3	–4	–4	–4	–4	–4	–4

11	12	13	14	15	16	17	18	19	20
4	3	6	7	4	8	9	5	3	6
4	5	–4	–4	3	–4	–4	–4	3	2
–4	–4	3	6	–4	3	1	7	–4	–4

21	22	23	24	25	26	27	28	29	30
8	9	2	7	4	6	9	8	3	2
–3	–1	6	–4	5	–4	–2	–4	4	5
–4	–4	–4	5	–4	7	–4	–1	–4	–4

Exercise 6.2: Minus 3 (– 3 = – 5 + 2)

1	2	3	4	5	6	7	8	9	10
6	1	7	9	2	4	7	9	8	9
2	3	2	–3	4	1	–2	–3	–1	–4
–3	–3	–3	2	–3	–3	–3	–3	–3	–3

11	12	13	14	15	16	17	18	19	20
3	3	6	7	3	8	9	5	4	6
3	5	–3	–3	4	–4	–3	–3	4	1
–3	–3	3	5	–3	–3	2	6	–3	–3

(Exercise 6.2 continued)

21	22	23	24	25	26	27	28	29	30
8	9	1	7	4	6	9	7	3	2
−3	−1	6	−3	5	−3	−2	−3	4	5
−3	−3	−3	5	−3	6	−3	−1	−3	−3

+-+

Exercise 6.3: Minus 2 (− 2 = − 5 + 3)

1	2	3	4	5	6	7	8	9	10
6	1	7	7	2	4	7	9	7	9
1	3	1	−2	4	1	−2	−3	−1	−4
−2	−2	−2	3	−2	−2	−2	−2	−2	−2

11	12	13	14	15	16	17	18	19	20
3	2	6	6	5	8	9	5	4	8
3	5	−2	−2	1	−4	−2	−2	4	−3
−2	−2	3	5	−2	−2	−2	6	−2	−2

21	22	23	24	25	26	27	28	29	30
9	9	1	8	2	5	5	6	3	3
−3	−1	5	−2	3	−2	−2	−2	2	5
−2	−2	−2	−2	−2	5	4	−4	−2	−2

Exercise 6.4: Minus 1 (– 1 = – 5 + 4)

1	2	3	4	5	6	7	8	9	10
1	3	4	8	2	2	5	9	8	6
1	−1	−1	−1	4	3	5	−4	−4	−1
−1	5	4	−3	−1	−1	−1	−1	−1	−1

11	12	13	14	15	16	17	18	19	20
9	4	8	10	2	5	10	1	1	3
1	1	−3	−1	7	−1	−5	4	9	−1
−1	−1	−1	1	−1	6	−1	−1	−1	8

+–+

Exercise 7.1: Add 9 (+ 9 = – 1 + 10)

1	2	3	4	5	6	7	8	9	10
2	1	3	4	5	4	2	4	1	5
2	3	9	9	3	5	3	2	9	9
9	9	5	2	9	9	9	9	7	4

11	12	13	14	15	16	17	18	19	20
4	8	5	8	9	7	7	6	5	8
−2	−5	9	9	−7	−3	9	9	−1	−6
9	9	−3	−5	9	9	−3	−2	9	9

21	22	23	24	25	26	27	28	29	30
1	9	3	6	5	7	2	4	8	9
9	9	9	9	9	9	9	9	9	5
9	9	9	9	9	9	9	9	9	9

Exercise 7.2: Add 8 (+ 8 = − 2 + 10)

1	2	3	4	5	6	7	8	9	10
2	1	3	4	5	4	2	4	1	5
2	3	9	8	3	5	3	2	8	8
8	8	8	3	8	8	8	8	3	4

11	12	13	14	15	16	17	18	19	20
4	8	5	8	8	7	7	6	5	8
−2	−5	8	8	−6	−3	8	8	−1	−4
8	8	−3	−5	8	8	−3	−2	8	8

21	22	23	24	25	26	27	28	29	30
1	8	3	6	5	7	2	4	9	5
8	8	8	8	8	8	8	8	8	9
8	8	8	8	8	8	8	8	8	8

Exercise 7.3: Add 7 (+ 7 = − 3 + 10)

1	2	3	4	5	6	7	8	9	10
2	1	3	4	5	4	2	4	1	5
2	3	7	7	3	5	3	2	7	7
7	7	8	3	7	7	7	7	3	7

11	12	13	14	15	16	17	18	19	20
4	8	5	7	7	7	8	6	5	8
−2	−5	7	7	−5	−3	7	7	−2	−4
7	7	−2	1	7	7	−3	−1	7	7

(Add 7 continued)

21	22	23	24	25	26	27	28	29	30
1	7	3	6	5	9	2	4	8	9
7	7	7	7	7	7	7	7	7	9
7	7	7	7	7	7	7	7	7	7

+-+

Exercise 7.4: Add 6 (+ 6 = − 4 + 10)

1	2	3	4	5	6	7	8	9	10
2	1	4	9	5	4	2	4	2	5
2	3	6	6	4	5	3	2	6	6
6	6	8	2	6	6	6	6	3	6

11	12	13	14	15	16	17	18	19	20
4	9	6	6	7	7	9	7	5	8
−1	−5	6	6	−5	−3	6	6	−1	−4
6	6	−2	7	6	6	−3	−1	6	6

21	22	23	24	25	26	27	28	29	30
1	6	3	5	9	2	4	7	8	9
6	6	6	6	6	6	6	6	6	8
6	6	6	6	6	6	6	6	6	6

Exercise 7.5: Miscellaneous problems: addition (mainly 5, 6, 7, 8, 9)

1	2	3	4	5	6	7	8	9	10
3	3	2	5	5	7	9	4	5	6
2	3	8	5	6	6	7	4	4	5
5	4	5	6	9	8	8	9	7	8

11	12	13	14	15	16	17	18	19	20
7	9	7	7	3	5	4	5	4	3
6	8	6	1	8	3	7	5	9	7
7	8	6	9	9	8	8	9	7	8

21	22	23	24	25	26	27	28	29	30
4	4	4	4	7	7	1	5	2	1
7	9	8	6	7	7	4	8	8	9
9	9	8	7	7	9	9	7	6	5
3	7	6	6	4	2	5	9	4	5

31	32	33	34	35	36	37	38	39	40
3	8	6	8	8	5	3	5	3	9
8	8	8	7	7	4	3	6	9	9
9	9	7	5	8	9	6	9	4	1
5	6	4	8	7	6	3	1	8	5

41	42	43	44	45	46	47	48	49	50
2	1	9	8	8	2	6	4	9	8
7	6	7	6	5	9	6	7	5	2
6	7	6	3	7	6	7	4	6	7
7	8	9	6	8	8	9	6	7	7
4	6	4	2	5	3	6	7	3	2

Exercise 7.6: Miscellaneous problems: addition/subtraction

1	2	3	4	5	6	7	8	9	10
3	3	9	9	5	7	9	5	5	8
2	3	−2	8	6	−3	8	−2	4	8
−5	−4	8	−6	−5	8	−4	7	−3	−3

11	12	13	14	15	16	17	18	19	20
8	9	7	7	6	5	9	8	4	6
7	8	−4	−1	9	−3	−6	−4	9	9
−1	−7	7	8	−5	9	7	8	−5	−1

21	22	23	24	25	26	27	28	29	30
8	9	4	4	7	7	1	5	6	1
−2	9	8	6	7	9	4	8	−4	9
4	−5	8	−1	−7	−2	9	3	6	−5
−2	6	−3	7	8	8	−5	−5	8	7

31	32	33	34	35	36	37	38	39	40
7	8	6	8	8	5	9	7	8	9
−3	8	−3	7	7	4	6	9	9	1
5	- 4	6	−3	−4	−8	−1	- 2	−3	−5
7	8	9	9	8	9	8	8	6	8

41	42	43	44	45	46	47	48	49	50
2	1	9	8	8	7	6	4	9	8
7	6	7	6	5	8	9	7	5	2
6	−3	−2	7	7	−2	−3	6	8	7
−3	9	7	5	6	8	7	−4	−2	−3
8	6	9	−3	−4	8	5	8	6	−3

Exercise 8.1: Add 4 with carry to left hand (+ 4 = − 6 + 10)

1	2	3	4	5	6	7	8	9	10
4	3	2	3	3	5	9	8	6	8
2	3	5	4	5	4	−2	4	−1	4
4	4	4	4	4	4	4	2	4	3

11	12	13	14	15	16	17	18	19	20
9	8	5	8	4	7	6	9	9	6
−1	−2	2	−1	4	−1	2	−3	−5	3
4	4	4	4	4	4	4	4	4	4

21	22	23	24	25	26	27	28	29	30
8	2	3	9	3	9	7	5	8	2
7	4	5	4	6	−2	7	3	−3	6
4	4	4	−2	4	4	4	4	4	4

Exercise 8.2: Add 3 with carry to left hand (+ 3 = − 7 + 10)

1	2	3	4	5	6	7	8	9	10
6	1	2	5	4	3	9	8	4	8
3	6	5	3	4	4	−1	1	3	3
3	3	3	3	3	3	3	3	3	4

11	12	13	14	15	16	17	18	19	20
9	8	5	8	7	7	6	8	9	6
3	−1	4	−2	3	2	2	3	3	2
3	3	3	3	8	3	3	1	−2	3

21	22	23	24	25	26	27	28	29	30
8	7	3	3	1	9	8	7	9	2
7	3	3	3	7	3	3	3	3	7
3	5	−5	3	3	6	7	7	5	3

Exercise 8.3: Add 2 with carry to left hand (+ 2 = − 8 + 10)

1	2	3	4	5	6	7	8	9	10
8	9	8	9	1	3	4	5	4	6
2	2	2	2	8	5	5	3	4	2
7	6	5	7	2	2	2	2	2	2

11	12	13	14	15	16	17	18	19	20
7	6	6	8	5	2	8	9	1	9
2	1	3	2	4	7	1	2	7	−1
2	2	2	6	2	2	2	4	2	2

21	22	23	24	25	26	27	28	29	30
8	9	3	7	2	9	8	9	2	8
2	2	6	1	6	2	2	2	9	2
9	5	2	2	2	8	8	9	2	4

+-+

Exercise 8.4: General exercises: single-digit

1	2	3	4	5	6	7	8	9	10
8	9	8	6	7	7	1	3	4	9
2	3	4	4	3	2	4	5	3	2
6	5	6	7	7	4	3	5	4	7
4	4	4	5	2	4	3	6	9	4

11	12	13	14	15	16	17	18	19	20
3	8	6	9	7	8	9	7	7	8
6	3	4	1	7	2	4	9	3	4
4	9	6	9	5	8	5	2	7	7
7	6	4	1	4	2	3	3	3	2

(Exercise 8.4 continued)

21	22	23	24	25	26	27	28	29	30
2	3	2	3	8	9	6	4	5	8
7	5	6	6	5	2	5	7	9	2
6	2	3	2	7	6	7	8	5	7
3	9	8	6	8	4	4	4	3	4
4	8	4	6	4	7	3	7	6	5

31	32	33	34	35	36	37	38	39	40
7	9	9	8	8	7	6	4	9	9
3	4	−2	3	4	8	9	6	3	2
8	5	4	7	7	−2	−3	8	8	6
−4	−6	7	2	8	8	7	−2	6	8
9	8	4	−2	4	8	5	4	−3	−2

Exercise 9.1: Minus 5 (− 5 = + 5 − 10)

1	2	3	4	5	6	7	8	9	10
17	12	16	15	11	18	13	14	10	19
−5	−5	−5	−5	−5	−5	−5	−5	−5	−5
2	2	3	6	6	8	4	5	7	−5

11	12	13	14	15	16	17	18	19	20
4	8	9	8	12	14	9	19	15	11
8	6	7	9	6	3	9	−3	−1	8
−5	−5	−5	−5	−5	−5	−5	−5	−5	−5

21	22	23	24	25	26	27	28	29	30
14	19	13	16	15	14	12	17	9	18
8	7	−5	8	6	−5	2	−5	6	−5
−5	−5	8	−5	−5	8	−5	8	−5	7

Exercise 9.2: Minus 9 (− 9 = + 1 − 10)

1	2	3	4	5	6	7	8	9	10
12	17	11	16	15	18	13	14	10	19
−9	−9	−9	−9	−9	−9	−9	−9	−9	−9
2	2	6	3	6	8	4	5	7	−9

11	12	13	14	15	16	17	18	19	20
4	8	9	8	12	14	9	19	15	11
8	5	5	9	6	3	7	−3	−1	8
−9	−9	−9	−9	−9	−9	−9	−9	−9	−9

21	22	23	24	25	26	27	28	29	30
14	19	13	16	15	14	12	17	9	18
8	7	−9	8	6	−9	5	−9	6	−9
−9	−9	8	−9	−9	8	−9	8	−9	7

Exercise 9.3: Minus 8 (− 8 = + 2 − 10)

1	2	3	4	5	6	7	8	9	10
12	17	11	16	15	18	13	14	10	19
−8	−8	−8	−8	−8	−8	−8	−8	−8	−8
2	2	6	3	6	8	4	5	7	−8

11	12	13	14	15	16	17	18	19	20
4	9	8	8	12	14	9	19	15	11
9	5	5	9	6	3	7	−3	−1	9
−8	−8	−8	−8	−8	−8	−8	−8	−8	−8

21	22	23	24	25	26	27	28	29	30
14	19	13	16	15	14	12	17	9	18
8	7	−8	8	6	−8	5	−8	6	−8
−8	−8	8	−8	−8	8	−8	8	−8	7

Exercise 9.4: Minus 7 (– 7 = + 3 – 10)

1	2	3	4	5	6	7	8	9	10
17	11	12	16	15	18	13	14	10	19
−7	−7	−7	−7	−7	−7	−7	−7	−7	−7
2	6	2	3	6	8	4	5	8	−7

11	12	13	14	15	16	17	18	19	20
4	8	9	8	12	14	9	19	15	11
8	5	5	9	6	3	7	−3	−1	8
−7	−7	−7	−7	−7	−7	−7	−7	−7	−7

21	22	23	24	25	26	27	28	29	30
14	19	13	16	15	14	12	17	9	18
8	7	−7	8	6	−7	5	−7	6	−7
−7	−7	8	−7	−7	8	−7	8	−7	6

Exercise 9.5: Minus 6 (– 6 = + 4 – 10)

1	2	3	4	5	6	7	8	9	10
17	16	18	11	15	12	13	14	10	19
−6	−6	−6	−6	−6	−6	−6	−6	−6	−6
2	3	8	6	6	2	4	5	7	−6

11	12	13	14	15	16	17	18	19	20
4	8	9	8	12	14	9	19	15	11
8	5	5	9	8	3	7	−3	−1	8
−6	−6	−6	−6	−6	−6	−6	−6	−6	−6

21	22	23	24	25	26	27	28	29	30
14	19	13	16	15	14	12	17	9	18
8	7	−6	8	7	−6	5	−6	6	−6
−6	−6	8	−6	−6	8	−6	8	−6	7

Exercise 9.6: Subtraction of numbers 5 to 9

1	2	3	4	5	6	7	8	9	10
16	14	19	13	21	17	20	26	15	11
−5	−5	−6	−6	−7	−9	−8	−7	−6	−5
−7	8	−5	8	−5	−4	−9	−5	−7	−2

11	12	13	14	15	16	17	18	19	20
18	15	23	17	24	19	20	18	22	16
−4	−7	−9	−8	−8	−5	−6	−7	−8	−2
−7	−4	−6	−6	−3	−6	−9	−7	−5	−8

21	22	23	24	25	26	27	28	29	30
8	9	4	9	16	17	15	22	23	21
9	9	8	7	7	8	8	−8	−8	−7
−6	−8	−5	−3	−9	−5	−6	7	6	8
−6	−8	−3	−6	−8	−7	−6	−9	−5	−6

31	32	33	34	35	36	37	38	39	40
24	18	28	27	18	15	19	26	23	22
−6	7	−9	−8	7	−6	−7	−8	−5	−6
2	−8	6	5	−5	7	6	5	7	7
−7	−5	−7	−6	−5	−8	−9	−8	−6	−8

41	42	43	44	45	46	47	48	49	50
9	8	11	20	17	16	26	14	19	18
8	7	−7	6	8	6	−3	7	5	2
−6	7	9	−7	−6	−8	7	−5	−8	−7
9	−5	8	6	9	7	−9	8	6	8
−7	−4	−5	−7	−7	−5	−8	−9	−7	−6

Exercise 10.1: Minus 4 (– 4 = + 6 – 10)

1	2	3	4	5	6	7	8	9	10
10	11	12	13	14	15	16	17	18	19
–4	–4	–4	–4	–4	–4	–4	–4	–4	–4
–4	–4	–4	–4	–4	–4	–4	–4	–4	–4

11	12	13	14	15	16	17	18	19	20
4	4	4	4	4	4	4	4	4	4
6	7	8	9	11	12	13	14	15	16
–4	–4	–4	–4	–4	–4	–4	–4	–4	–4

21	22	23	24	25	26	27	28	29	30
14	19	13	16	15	11	12	17	9	14
8	7	–4	8	6	–4	5	–4	6	7
–4	–4	8	–4	–4	8	–4	7	–4	–4

Exercise 10.2: Minus 3 (– 3 = + 7 – 10)

1	2	3	4	5	6	7	8	9	10
10	11	12	13	14	15	16	17	18	19
–3	–3	–3	–3	–3	–3	–3	–3	–3	–3
–3	–3	–3	–3	–3	–3	–3	–3	–3	–3

11	12	13	14	15	16	17	18	19	20
3	3	3	3	3	3	3	3	3	3
6	7	8	9	11	12	13	14	15	16
–3	–3	–3	–3	–3	–3	–3	–3	–3	–3

21	22	23	24	25	26	27	28	29	30
14	19	10	16	15	11	12	17	9	14
8	7	–3	8	6	–3	5	–3	6	7
–3	–3	8	–3	–3	8	–3	7	–3	–3

Exercise 10.3: Minus 2 (− 2 = + 8 − 10)

1	2	3	4	5	6	7	8	9	10
10	11	12	13	14	15	16	17	18	19
−2	−2	−2	−2	−2	−2	−2	−2	−2	−2
−2	−2	−2	−2	−2	−2	−2	−2	−2	−2

11	12	13	14	15	16	17	18	19	20
2	2	2	2	2	2	2	2	2	2
6	7	8	9	11	12	13	14	15	16
−2	−2	−2	−2	−2	−2	−2	−2	−2	−2

21	22	23	24	25	26	27	28	29	30
13	19	10	16	15	11	12	16	9	14
8	7	−2	8	5	−2	5	−2	6	7
−2	−2	8	−2	−2	8	−2	7	−2	−2

+-+

Exercise 10.4: Miscellaneous (minus 2, 3, 4)

1	2	3	4	5	6	7	8	9	10
8	7	3	5	6	9	5	7	8	8
3	5	9	8	6	7	−2	−3	−4	7
−4	−3	−2	−4	−3	−3	8	7	9	−4
9	6	7	3	−5	−4	−4	−2	−4	−3

11	12	13	14	15	16	17	18	19	20
12	22	10	13	21	15	14	13	32	31
−4	−3	−2	−4	−3	−3	9	8	−4	−3
8	2	−3	6	−4	−3	−4	−2	3	2
−3	−4	12	−2	6	2	8	3	−2	−3

Exercise 11.1: General addition of 2-digit numbers

1	2	3	4	5	6	7	8	9	10
13	23	52	35	12	73	27	14	63	76
25	21	31	54	60	25	21	54	26	12

11	12	13	14	15	16	17	18	19	20
17	19	16	23	27	35	24	52	44	29
15	19	19	17	28	55	57	27	38	57

21	22	23	24	25	26	27	28	29	30
43	54	36	45	35	37	18	25	12	39
48	39	27	27	28	37	26	38	48	28

31	32	33	34	35	36	37	38	39	40
48	56	39	38	45	59	28	57	68	45
23	35	22	46	39	24	53	38	16	38

41	42	43	44	45	46	47	48	49	50
36	77	33	28	35	36	29	78	55	49
37	17	47	66	46	47	63	13	28	47

51	52	53	54	55	56	57	58	59	60
11	12	13	15	17	18	14	28	25	15
12	13	14	16	18	16	19	13	18	47
13	14	15	17	19	16	18	26	28	29

61	62	63	64	65	66	67	68	69	70
36	23	19	57	28	49	25	26	43	32
14	34	37	16	43	17	26	38	26	43
28	38	25	24	19	18	27	17	28	16

Exercise 11.2: General subtraction of 2-digit numbers

1	2	3	4	5	6	7	8	9	10
23	33	41	48	35	46	52	70	53	44
-15	-19	-27	-17	-18	-29	-24	-16	-28	-25

11	12	13	14	15	16	17	18	19	20
64	67	56	83	71	90	95	79	77	53
-35	-28	-27	-15	-24	-34	-39	-26	-29	-18

21	22	23	24	25	26	27	28	29	30
64	30	66	82	81	72	35	77	55	93
-36	-19	-38	-17	-44	-57	-16	-39	-38	-26

31	32	33	34	35	36	37	38	39	40
36	42	61	94	40	65	87	50	94	99
-18	-26	-33	-48	-25	-47	-38	-21	-47	-48

41	42	43	44	45	46	47	48	49	50
36	23	29	18	28	39	25	66	43	62
14	38	37	16	53	24	26	-28	-26	-43
-28	-24	-27	-19	-22	-18	-27	12	38	26

51	52	53	54	55	56	57	58	59	60
36	83	19	47	60	69	36	26	92	55
-9	-36	37	-16	-43	18	37	-17	-46	28
-12	27	-25	59	48	-39	-45	68	23	-37

Exercise 12.1: Multiplication 1 x 1

Do the following by repeated addition. For example, to do 5 x 3 (five 3s), you must add 3 while counting to 5.

1. 3 x 2	2. 3 x 3	3. 2 x 3	4. 2 x 4	5. 2 x 5
6. 4 x 2	7. 4 x 3	8. 5 x 2	9. 3 x 4	10. 6 x 2
11. 3 x 5	12. 7 x 2	13. 2 x 6	14. 5 x 3	15. 2 x 2
16. 8 x 2	17. 2 x 7	18. 6 x 3	19. 2 x 8	20. 7 x 3
21. 8 x 3	22. 3 x 7	23. 3 x 8	24. 9 x 2	25. 9 x 3
26. 3 x 6	27. 2 x 9	28. 3 x 9	29. 5 x 4	30. 6 x 7
31. 5 x 5	32. 6 x 8	33. 4 x 4	34. 5 x 8	35. 4 x 5
36. 6 x 9	37. 4 x 6	38. 7 x 4	39. 5 x 6	40. 7 x 7
41. 5 x 9	42. 7 x 8	43. 4 x 8	44. 5 x 7	45. 8 x 5
46. 8 x 4	47. 6 x 6	48. 9 x 5	49. 8 x 9	50. 4 x 7
51. 4 x 9	52. 6 x 5	53. 9 x 4	54. 7 x 5	55. 8 x 6
56. 6 x 4	57. 8 x 8	58. 7 x 9	59. 9 x 6	60. 8 x 7
61. 7 x 6	62. 9 x 7	63. 9 x 8	64. 9 x 9	

+‑+

Exercise 12.2: Multiplication 2 x 1

1. 86 x 2	2. 48 x 3	3. 38 x 4	4. 53 x 5	5. 26 x 6
6. 92 x 7	7. 41 x 8	8. 28 x 2	9. 61 x 9	10. 33 x 3
11. 47 x 4	12. 64 x 5	13. 99 x 6	14. 44 x 7	15. 26 x 8
16. 35 x 9	17. 70 x 2	18. 76 x 3	19. 81 x 4	20. 97 x 5
21. 69 x 6	22. 77 x 7	23. 13 x 8	24. 29 x 9	25. 98 x 2
26. 23 x 3	27. 59 x 4	28. 62 x 5	29. 66 x 6	30. 46 x 7
31. 57 x 8	32. 16 x 9	33. 34 x 2	34. 83 x 3	35. 89 x 4
36. 21 x 5	37. 72 x 6	38. 94 x 7	39. 88 x 8	40. 67 x 9
41. 54 x 2	42. 52 x 3	43. 96 x 4	44. 39 x 5	45. 38 x 6
46. 78 x 7	47. 56 x 8	48. 49 x 9	49. 75 x 7	50. 68 x 8

Made in the USA
Las Vegas, NV
22 July 2021